How We Understand Others

In our everyday social interactions, we try to make sense of what people are thinking, why they act as they do, and what they are likely to do next. This process is called mindreading. Mindreading, Shannon Spaulding argues in this book, is central to our ability to understand and interact with others. Philosophers and cognitive scientists have converged on the idea that mindreading involves theorizing about and simulating others' mental states. She argues that this view of mindreading is limiting and outdated. Most contemporary views of mindreading vastly underrepresent the diversity and complexity of mindreading. She articulates a new theory of mindreading that takes into account cutting-edge philosophical and empirical research on in-group/out-group dynamics, social biases, and how our goals and the situational context influence how we interpret others' behavior.

Spaulding's resulting theory of mindreading provides a more accurate, comprehensive, and perhaps pessimistic view of our abilities to understand others, with important epistemological and ethical implications. Deciding who is trustworthy, knowledgeable, and competent are epistemically and ethically fraught judgments; her new theory of mindreading sheds light on how these judgments are made and the conditions under which they are unreliable.

This book will be of great interest to students of philosophy of psychology, philosophy of mind, applied epistemology, cognitive science and moral psychology, as well as those interested in conceptual issues in psychology.

Shannon Spaulding is Assistant Professor of Philosophy at Oklahoma State University, USA.

Routledge Focus on Philosophy

Routledge Focus on Philosophy is an exciting and innovative new series, capturing and disseminating some of the best and most exciting new research in philosophy in short book form. Peer reviewed and at a maximum of fifty thousand words, shorter than the typical research monograph, *Routledge Focus on Philosophy* titles are available in both ebook and print on demand format. Tackling big topics in a digestible format, the series opens up important philosophical research for a wider audience, and as such is invaluable reading for the scholar, researcher, and student seeking to keep their finger on the pulse of the discipline. The series also reflects the growing interdisciplinarity within philosophy and will be of interest to those in related disciplines across the humanities and social sciences.

Available:

- **Plant Minds** *Chauncey Maher*
- **The Logic of Commitment** *Gary Chartier*
- **Consciousness and Moral Status** *Joshua Shepherd*
- **The Passing of Temporal Well-Being** *Ben Bramble*
- **How We Understand Others** *Shannon Spaulding*
- **Consciousness and Moral Status** *Joshua Shepherd*

For more information about this series, please visit: www.routledge.com/Routledge-Focus-on-Philosophy/book-series/RFP

How We Understand Others

Philosophy and Social Cognition

Shannon Spaulding

Routledge
Taylor & Francis Group
LONDON AND NEW YORK

First published 2018
by Routledge
2 Park Square, Milton Park, Abingdon, Oxon OX14 4RN

and by Routledge
711 Third Avenue, New York, NY 10017

Routledge is an imprint of the Taylor & Francis Group, an informa business

British Library Cataloguing-in-Publication Data
A catalogue record for this book is available from the British Library

Library of Congress Cataloging-in-Publication Data
A catalog record for this book has been requested

ISBN: 978-1-138-22158-1 (hbk)
ISBN: 978-1-315-39606-4 (ebk)

Typeset in Times New Roman
by Apex CoVantage, LLC

For Leila, my joyful, silly, clever daughter.

Contents

Acknowledgments ix

1 **Introduction** 1

2 **In defense of mindreading** 7

2.1 Introduction to mindreading 7
2.2 The embodied and enactive cognition challenge 9
2.3 The pluralistic folk psychology challenge 15
2.4 Assessment 19

3 **Expanding the scope of mindreading: social categorization, stereotypes, social bias, and situational context** 23

3.1 The narrow focus of standard mindreading theories 23
3.2 Social categorization 25
3.3 Stereotypes 26
3.4 Social biases 27
3.5 Situational context 34
3.6 Assessment 35

4 **The goals of mindreading** 42

4.1 Why we mindread 42
4.2 The various goals of mindreading 43
4.3 Approaches to mindreading 45
4.4 Mindreading products 52
4.5 Assessment 56

5 Model theory 62

5.1 The function of theories 62
5.2 General theories of mindreading 64
5.3 Model theory 68
5.4 Assessment 72

6 Epistemic and ethical applications 80

6.1 The isolation of the mindreading literature 80
6.2 Judging others' knowledge and competence 81
6.3 Application to the epistemology of peer disagreement 85
6.4 Application to epistemic justice 88
6.5 Assessment 93

7 Concluding thoughts 98

Index 103

Acknowledgments

I am grateful to the editorial staff at Routledge for inviting me to write this book and for seeing it through from conception to completion. This book has been several years in the making. The inspiration for the broader conception of mindreading at the heart of this book came during a visiting research fellowship at the University of Edinburgh in 2015, where conversations with Suilin Lavelle proved incredibly helpful in crystalizing my ideas. Thanks both to the University of Edinburgh for hosting my fellowship and to the Oklahoma State University philosophy department for allowing me to take a research leave to work on this project. Though many people have helped me work through various ideas in this book, I am especially indebted to Guillermo del Pinal, Evan Westra, Tad Zawidzki, and Robert Thompson. Not only do I find inspiration in their work, these philosophers have read many drafts of my papers, and we have had numerous helpful conversations on the topics in this book. I am lucky to have such friendly and supportive colleagues in the field. I am also grateful to the anonymous reviewers of this book for their detailed, constructive feedback on the book manuscript. Finally, I owe many thanks to my husband Dave for taking on more than his share of the parenting so I could complete this book.

1 Introduction

Human beings are highly social creatures. We live in families, surround ourselves with friends, collaborate with colleagues, compete with rivals, and are interested in sports, politics, and celebrities we have never met. Our work lives are organized into social groups – companies, teams, universities, departments – and in our home life we care for, worry about, and spend time with friends and family. Even when we are alone, we think about other people and how other people think of us, read books and watch shows about real and fictional characters, and spend countless hours on social media. Sociality permeates and shapes every aspect of our lives.

Philosophers, psychologists, neuroscientists, linguists, and anthropologists study the cognitive capacities that underlie such social interactions. The capacity to understand and interact with other agents is called *folk psychology*. In philosophy and the cognitive sciences, the orthodox view[1] is that folk psychology is partly underwritten by *theory of mind*, or *mindreading*, the capacity to make sense of intentional behavior in terms of mental states. According to the orthodox view, successful social interaction often involves understanding what others are thinking and what they are trying to achieve. In interacting with colleagues, rivals, and loved ones, reading fiction, driving on a busy freeway, and many other of our daily social activities, we attribute beliefs, desires, emotions, and intentions to people to make sense of their behavior, and on the basis of that we predict what they are likely to do next. On this view, understanding another's behavior is achieved by generating mentalistic explanations and predictions.[2]

Those who subscribe to the orthodox view have offered different general accounts of mindreading. The Theory Theory (TT) constitutes one broad family of theories of mindreading. The TT is committed to the idea that mindreading is best understood as a kind of theorizing. In understanding others' behavior, we gather information, employ rich bodies of information about folk psychology, draw inferences about the unobserved causes of the behavior (i.e., the agent's mental states that led to that behavior), and on this basis make predictions about what the agent is likely to do in the future. In the early days, theorists conceived of mindreading as very closely analogous to scientific theorizing (Gopnik and Wellman 1992). Nowadays, there are many versions of the TT. For instance, some versions of the TT adhere to a modular view of the mindreading capacity (and within this camp, some theories view mindreading as strongly modular (Fodor 1983) and others view it as weakly modular (Carruthers 2006)), nativist theories (Scholl and Leslie 1999), constructivist theories (Wellman 2015), and theories based on Bayesian models of learning (Baker, Saxe, and Tenenbaum 2011). Though this family of views is quite diverse, they are unified in viewing mindreading as an information-rich inferential cognitive process.

Some other general theories of mindreading fall under the banner of the Simulation Theory (ST). Like versions of the TT, the theories in this category are diverse, but they share a few core commitments. All versions of the ST hold that the mindreading is based on a practical ability to take on a target's mental states, i.e., to mentally simulate a target, rather than draw inferences from a rich body of folk psychological information. There is a healthy debate amongst simulation theorist over the concept of simulation, with some arguing that simulation implies resemblance and others arguing that it implies reuse (Barlassina and Gordon 2017, Section 2). Introspective versions of the ST hold that we imagine ourselves in a target's situation, figure out what we would think, feel, and do in that situation, and then attribute that to the target (Goldman 2006, Ch. 7), whereas non-introspective versions of ST involve no introspection at all (Gordon 2007). Some versions of the ST focus only on the psychological implementation of imagination-based "high-level" simulational mindreading (Currie and Ravenscroft 2002), whereas others also focus on the neurological implementation of "low-level" simulational mindreading (Goldman 2009). Vigorous debate about these and other aspects of the ST continues, and just like the TT, contemporary versions of the ST are much more sophisticated and nuanced than the initial versions proposed over three decades ago.

Despite their differences, the various versions of the TT and the ST all adhere to the orthodox view of mindreading described above. The orthodox view is not without challenge, of course. Many theorists on the margins of the mainstream view and theorists squarely outside of mainstream philosophy and cognitive science object to the idea that our social interactions often are mediated by mental state attributions. For example, some philosophers argue that although we *can* mindread, we navigate most of our social interactions using psychological concepts other than belief and desire, e.g., personality traits, stereotypes, behavioral scripts etc. On this kind of view, we employ lots of different kinds of psychological concepts in our social practices, and the function of our social practices often is to regulate our own and others' behavior rather than explain and predict others' behavior (Andrews 2012; McGeer 2007; Zawidzki 2013). Even more radically, theorists from the embodied and enactive cognition research programs challenge the idea that we attribute any sort of mental states and explain and predict behavior in our ordinary social interactions. Embodied and enactive cognition theorists offer non-mentalistic accounts of social interaction that are meant to challenge and replace mindreading theories (Gallagher 2005, 2012; Hutto 2008).

In this book, I will argue that mindreading is an important tool in our folk psychological toolkit. By defending this view, I am in one sense articulating and defending a view that is squarely within the mainstream view. My arguments in this book, however, will challenge several aspects of the mainstream philosophical view of mindreading. In particular, I will argue that mindreading is not as simple, uniform, or accurate as the philosophical literature on mindreading suggests. Individuals differ with respect to informational input to mindreading, their goals in mindreading, the kind of mindreading strategies they adopt, and the kind of mindreading output they produce. My claim is not simply that individuals use their mindreading judgments differently. That much is uncontroversial. Rather, my claim is that the input, processing, and output of mindreading all vary along many dimensions, which makes constructing an empirically adequate account of mindreading significantly more challenging than typically recognized. The philosophical literature on mindreading suggests that neurotypical adult humans rarely make mindreading mistakes, that competent mindreaders all pretty much agree on the mentalistic explanations and predictions we infer, and all there really is to mindreading is attributing a belief, desire, or intention and explaining and predicting behavior. I will challenge each of these ideas in the following chapters. The overarching theme of this book is that mindreading is much more complex, messy, interesting, and relevant to other debates than philosophers have acknowledged.

Here is the outline for the rest of this book. In Chapter 2, I consider challenges to the idea that mindreading is a central and frequently used tool in our folk psychological toolkit. I argue that although aspects of these challenges are compelling, they fail to show that mindreading is a marginal, rarely used social practice. Nevertheless, they do highlight aspects of social interaction that have received little attention in the mindreading literature, e.g., the role of behavioral schemata and norms in social interactions.

In Chapter 3, I argue that existing mindreading theories often are too narrowly focused on explaining the attribution of beliefs. As a result, they neglect other aspects of social interaction, such as how social categorization, stereotypes, social biases, and situational context influence how we interpret social behavior. These other aspects often are opaque to mindreaders, but they filter the available information that serves as input to mindreading and thus directly influence the mental representations mindreaders end up attributing.

In Chapter 4, I consider how the various goals we have in a social interaction affect the process and product of mindreading. Most contemporary mindreading theories presuppose that our primary goal in mindreading is accuracy. I argue that while this is the case in certain conditions, in other contexts our primary goals are efficiency, confirming our preexisting ideas, validating our self-worth, manipulating others, etc. These different goals bring about various strategies for mindreading and generate different types of explanations of social behavior, which are relatively unexplored by current theories of mindreading. I discuss how our goals influence stereotype activation, the mindreading approaches we employ in social interactions, how differently we approach mindreading in-group and out-group members, and how reliable these various strategies are.

In Chapter 5, I evaluate existing theories of mindreading in light of the phenomena considered in Chapters 3 and 4. Although the main theories – hybrid versions of Theory Theory and Simulation Theory – are good insofar as they go, I argue that they are incomplete. We need a comprehensive theory of mindreading that also encompasses the broader set of phenomena I canvass in earlier chapters. In this chapter, I articulate and defend Model Theory, which has the potential to be the kind of comprehensive, unifying theory of mindreading that we need.

In Chapter 6, I apply the lessons from my discussion of mindreading to debates outside of philosophy of psychology. Most philosophical discussions of mindreading are pretty isolated from other areas of philosophy, such as ethics and epistemology. I think this is a mistake, because understanding how we interpret and interact with others is important for other areas of philosophy, as well. I consider debates in social epistemology, feminist philosophy, and ethics that concern judging others as knowledgeable and competent with respect to some issue. These judgments clearly involve mindreading, and the updated theory of mindreading I articulate in the previous chapter can help shed light on how these judgments occur, when and how they are biased, and when not to trust these judgments. This discussion is important for understanding our judgments of who is an epistemic peer, analyzing disagreements amongst people we judge to be epistemic peers, epistemic injustice, and theorizing about interventions to prevent epistemic injustice.

I conclude with Chapter 7, where I take a birds-eye view of the perspective on mindreading presented in this book. The more comprehensive view of mindreading I describe in the book is somewhat cynical, but I argue that it is justified both empirically and rhetorically. In addition, I describe several exciting areas of research for mindreading theorists going forward. These include constructing, elaborating, and testing other general theories of mindreading that are genuinely comprehensive, investigating the developmental trajectories and interaction between mindreading and social biases, applying this comprehensive view of mindreading to the study of Autism Spectrum Disorder, and accounting for the variety of explanatory practices in mindreading.

Notes

1 By *orthodox* or *mainstream* view, I mean that most of the philosophical and empirical work adheres to this view and the most influential works in the field are expressly committed to this view. Indeed, despite vigorous challenges from enactive and embodied cognition, many people who subscribe to the orthodox view of mindreading do not regard it as at all controversial.

2 Though we can explain and predict our own behavior – and there are vigorous debates about whether or not this capacity is underwritten by the same capacities we use to explain and predict others' behavior – in this book for the most part I will focus on third-person rather than first-person explanation and prediction.

References

Andrews, K. 2012. *Do Apes Read Minds? Toward a New Folk Psychology*. Cambridge, MA: MIT Press.

Baker, C., R. Saxe, and J. Tenenbaum. 2011. "Bayesian theory of mind: Modeling joint belief-desire attribution." Proceedings of the Annual Meeting of the Cognitive Science Society.

Barlassina, L., and R. M. Gordon. 2017. "Folk psychology as mental simulation." In *Stanford Encyclopedia of Philosophy*, edited by E. N. Zalta. Stanford: Stanford University Press.

Carruthers, P. 2006. *The Architecture of the Mind*. Oxford: Oxford University Press.

Currie, G., and I. Ravenscroft. 2002. *Recreative Minds: Imagination in Philosophy and Psychology*. Oxford: Oxford University Press.

Fodor, J. A. 1983. *The Modularity of Mind*. Cambridge, MA: MIT Press.

Gallagher, S. 2005. *How the Body Shapes the Mind*. New York: Oxford University Press.

Gallagher, S. 2012. "In defense of phenomenological approaches to social cognition: Interacting with the critics." *Review of Philosophy and Psychology* 3 (2): 187–212.

Goldman, A. I. 2006. *Simulating Minds: The Philosophy, Psychology, and Neuroscience of Mindreading (Philosophy of Mind)*. New York: Oxford University Press.

Goldman, A. I. 2009. "Mirroring, simulating and mindreading." *Mind & Language* 24 (2): 235–252.

Gopnik, A., and H. M. Wellman. 1992. "Why the child's theory of mind really is a theory." *Mind & Language* 7 (1–2): 145–171.

Gordon, R. M. 2007. "Ascent routines for propositional attitudes." *Synthese* 159 (2): 151–165.

Hutto, D. D. 2008. *Folk Psychological Narratives: The Sociocultural Basis of Understanding Reasons*. Cambridge, MA: MIT Press.

McGeer, V. 2007. "The regulative dimension of folk psychology." In *Folk Psychology Re-assessed*, 137–156. New York: Springer.

Scholl, B. J., and A. M. Leslie. 1999. "Modularity, development and 'theory of mind'." *Mind & Language* 14 (1): 131–153.

Wellman, H. M. 2015. *Making Minds: How Theory of Mind Develops*. Oxford: Oxford University Press.

Zawidzki, T. W. 2013. *Mindshaping: A New Framework for Understanding Human Social Cognition*. Cambridge, MA: MIT Press.

2 In defense of mindreading

Mindreading, as it is typically conceived, consists of attributing beliefs, desires, and other mental states to an agent in order to explain and predict that agent's behavior. In this chapter, I explain the mainstream view of mindreading and some of the most vigorous challenges to the mainstream view. I consider objections to mindreading from embodied and enactive cognition and from pluralistic folk psychology. I argue that both sorts of objections fail to show that mindreading is unnecessary, rare, or unimportant. However, the challenge from pluralistic folk psychology highlights the fact that there is much more to mindreading than typically recognized by mindreading theories.

2.1 Introduction to mindreading[1]

The standard mindreading story is that in order to understand and successfully interact with other agents, neurotypical adult humans attribute mental states to other agents in order to explain and predict their behavior.[2] Neurotypical adult humans tend to find human behavior intrinsically interesting and want to understand why people behave the way they do. This is evident in the consumption of fiction, reality TV, celebrity gossip, politics, and the myriad examples of our speculating about others' motives and intentions. In addition to finding human behavior interesting, we explain others' behavior in order to make predictions about what they will do in the future, to coordinate our own actions with theirs, and perhaps to control and manipulate their behavior. (See Table 2.1 for a description of the terminology used in these debates.)

The following example is the kind of case that mindreading theorists would posit as an explanation of how mindreading works. Suppose Lawrence agrees to meet you at your office at 11 a.m., and by 11:10 a.m. he has not come to your office. The standard mindreading story holds that we are inclined to explain Lawrence's behavior by figuring out what his beliefs

Table 2.1 Standard Terminology

Social Cognition: Cognitive processes underlying interactions with other people
– Attributional and non-attributional processes cognitive processes in social interactions, e.g., attention, memory, and affect

Folk Psychology: Cognitive capacity to make sense of others' behavior

Theory of Mind: Cognitive capacity to infer mental states in order to make sense of others' behavior
– Attributing beliefs, desires, intentions, and emotions to others
– Typically the function is to explain and predict behavior
– Synonymous with "mindreading"

Mindreading: Cognitive capacity to infer mental states in order to make sense of others' behavior
– Attributing beliefs, desires, intentions, and emotions to others
– Typically the function is to explain and predict behavior
– Synonymous with "Theory of Mind"
– Preferred term because it omits the Theory Theory connotation implicit in "Theory of Mind"

and desires are. We might consider the possibility that Lawrence desires to meet with you but has a false belief about the meeting time or location. Or perhaps Lawrence desires to meet with you, has a true belief about the scheduled meeting time and location, but something urgent has come up and he cannot come to your office. Or perhaps Lawrence has a true belief about the scheduled meeting time and location but does not desire to meet with you. Through a process of theorizing or simulating, or a combination of both, you determine which of these belief/desire combinations is most plausible and attribute them to Lawrence. This explanation allows you to predict Lawrence's future behavior – perhaps he will show up late, send an apologetic email after the urgent situation is resolved, or simply avoid contacting you altogether.[3]

The standard mindreading story is that this process of attributing mental states to explain and predict behavior happens all the time in ordinary social interactions. Inferring mental states, explaining, and predicting behavior may occur consciously, but of course they need not. As I explained in the first chapter, all versions of the TT, the ST, and hybrid accounts subscribe to this standard mindreading story. Though these theories differ with respect to the mechanics of mindreading, they all *agree* that how we understand and interact with others in social environments is by explaining and predicting their behavior on the basis of mental state attributions. In the mindreading literature, it is often maintained that mindreading is the primary, pervasive way we understand others, that mindreading is essential for navigating the social world, and that we could not understand and interact in our social environments as successfully and easily as we do if we were not avid mindreaders. I call this *the broad scope of mindreading claim* (Spaulding 2010, 122).

Although the broad scope claim traditionally has been widely accepted in the mainstream empirical and philosophical mindreading literature (Baron-Cohen 1995, 3; Leslie 2000, 1236; Tooby and Cosmides 1995, xvii; Carruthers 2009; Goldman 2006, 3; Fodor 1987), recently theorists outside, on the margins, and even within the mainstream mindreading literature challenge this claim. I shall argue that some of these challenges are misguided, some are inconclusive, and others are compelling and pave the way for a better account of mindreading. I will consider the most radical challenges first.

2.2 The embodied and enactive cognition challenge

The first and most radical challenge to the broad scope claim comes from a family of views in the embodied and enactive cognition camp. This is a diverse family of theories, with nuanced and substantial distinctions amongst the different versions of each view. However, one uniting element of embodied and enactive cognition is the rejection the cognitivism. Cognitivism holds that our cognitive capacities are underwritten by computational procedures operating on symbolic, internal mental states, and cognitive science should study these internal states and processes. Embodied and enactive cognition – EC for short – holds that cognitivism misguidedly views the mind as something to be studied independently of the body and its environment. Cognitive science ought to investigate how the body and the environment shape the mind (Gallagher 2017). EC theorists in the folk psychology subfield apply this general critique of cognitivism to the idea that our folk psychological practices are underwritten by mental state attributions.

EC theorists argue that folk psychology is not based on mental state attributions. To the contrary, basic, non-mentalistic embodied practices are what underlie our ordinary social interactions. These embodied practices are constituted by "primary intersubjectivity" and "secondary intersubjectivity." Primary intersubjectivity, on this view, is the pre-theoretical, non-conceptual, embodied understanding of others that underlies and supports the higher-level cognitive skills posited in the mindreading literature. Shaun Gallagher characterizes it as the "innate or early developing capacity to interact with others manifested at the level of perceptual experience – we see or more generally perceive in the other person's bodily movements, facial gestures, eye direction, and so on, what they intend and what they feel" (Gallagher 2004, 204). Primary intersubjectivity manifests as the capacity for facial imitation and proprioceptive sense of one's body, the capacity to detect and track eye movement, to detect intentional behavior, and to read emotions from actions and expressive movements of others (Gallagher 2017, 155–158). Primary intersubjectivity is described in EC terminology here, but everyone in these debates should agree that these capacities are necessary for the development of mindreading.

Primary intersubjectivity is an essential element of folk psychology, but it is only with the addition of secondary intersubjectivity that we develop sophisticated folk psychological abilities. Secondary intersubjectivity emerges around age 1, and it is marked by a move from one-on-one, immediate intersubjectivity to contexts of shared attention. In addition to tracking eye movement, detecting intentional behavior, and reading emotions, with the development of secondary intersubjectivity, the child develops the capacity to engage in shared attention behaviors. As Peter Hobson describes it, "The defining feature of secondary intersubjectivity is that an object or event can become a focus *between* people. Objects and events can be communicated about . . . the infant's interactions with another person begin to have reference to the things that surround them" (Hobson 2004, 62). Following gazes, pointing, and communicating with others about objects of shared attention helps further develop the child's capacity for social understanding. Mindreading proponents and EC theorists both agree on the importance of this development, but they disagree on how to interpret it. According to EC, joint attention is behavioral and non-mentalistic, whereas mindreading theories hold that getting others to attend to an object and communicating with them about that object involve inferences about others' mental states.

According to EC, developing the capacity for mindreading requires mastering primary and secondary intersubjectivity, which give meaning to intersubjective interactions.[4] On this view, embodied practices are developmentally fundamental. Though there are vigorous disagreements about the best way to interpret EC's arguments about the development of folk psychology, I will not delve into these debates here. See Spaulding (2010, 2011) for more on these specific arguments. I shall instead focus on the more controversial EC claim that, even as adults, our ordinary interactions rarely go beyond primary and secondary intersubjectivity. According to EC, only in very unusual circumstances do we attribute mental states to others and explain and predict their behavior on the basis of those attributed mental states. On this view, we mindread only when we encounter *inexplicable* behavior. In the vast majority of social interactions, we get by with non-mentalistic interpretations of embodied cues. As Gallagher says, "In most of our ordinary and everyday intersubjective situations we have a direct, perception-based understanding of another person's intentions because their intentions are explicitly expressed in their embodied actions" (2004, 205).

Mindreading theorists are not committed to the claim that mindreading is the *only* way by which we understand others. Not every instance of socializing is underpinned by mental state attributions and explanation and prediction.[5] Thus, evaluating the broad scope claim requires examining whether, as mindreading theories posit, mindreading is common in ordinary circumstances or, as EC theories maintain, mindreading is *very* rare. If EC can establish that we engage in mindreading *only* in the rare circumstances in which we encounter inexplicable behavior, this would validate EC's claim that the mainstream literature on folk psychology has been massively misguided.

So, what are the arguments regarding the prevalence of mindreading? Shaun Gallagher offers the most well-developed EC arguments against mindreading. I will consider his most influential arguments here. Gallagher argues that there is good phenomenological evidence against the claim that mindreading is our primary and pervasive mode of understanding others. Gallagher argues that careful phenomenological reflection shows that our ordinary interactions do not seem to involve mental state inferences or explanation and prediction (Gallagher 2001, 89–92; 2004, 201–204; 2005, 208–216).

One could object, of course, that much of mindreading is supposed to occur subconsciously, especially in cases where the situation is fairly typical, and phenomenology cannot tell us whether mindreading occurs subconsciously. Gallagher acknowledges this but argues that

> a careful and methodical phenomenology should be able to tell us whether, when we hear the exasperated voice, our usual response involves formulating an explanation or predicting what the person will do next. Our encounters with others are in fact not normally occasions for theorizing or simulating if such nonconscious procedures are cashed out phenomenologically as explaining or predicting on the basis of postulated mental states.
>
> (Gallagher 2001, 89)

In other words, Gallagher argues that the mental state inferences posited by mindreading may be subconscious but the product, explanation or prediction, would be conscious and phenomenologically assessable. Careful introspection reveals there is no phenomenological evidence of explanation and prediction in typical social interactions. Our interactions are best characterized as immediate, evaluative, and phenomenologically fluid. Even when we are adopting a third-person stance (e.g., when we are watching but not taking part in a conversation), there is no phenomenological evidence of explanation and prediction. Thus, Gallagher concludes, tacit mindreading is not going on in ordinary social interactions. It is only when we are in an unfamiliar situation, or when we encounter very odd behavior, that we explain and predict behavior (Gallagher 2005, 208–215). If he is right, then the broad scope of mindreading claims is false.

This debate hinges on whether phenomenological evidence can play the kind of role that Gallagher proposes, i.e., whether it can serve as the basis for decisively rejecting or accepting theories of folk psychology. We can distinguish a weaker and a stronger methodological role for phenomenology in these debates.[6] Phenomenology may weakly constrain theories of folk psychology in the sense that theories of folk psychology ought to be compatible with our phenomenological experiences, or phenomenology may strongly constrain folk psychological theories in the sense that theories ought to explain and be empirically supported by our phenomenological experiences. I distinguish these two methodological roles because when mindreading theorists reject phenomenology as irrelevant, this is often interpreted as rejecting something like the weak constraint (Gallagher 2012). But, of course, one can accept that theories ought not make false claims about our phenomenology without accepting theories ought to explain and be empirically supported by our phenomenological experience.

Mindreading theorists and EC theorists both ought to accept that phenomenology *weakly* constrains the study of folk psychology. That is, they ought to agree that it counts against a theory if it entails a phenomenology radically different from what we experience. This is a very weak constraint that, as far as I can tell, none of the theories being debated violate. Theories that have *no* implications for our phenomenological experience of social interactions are compatible with the weak constraint. If the claim at issue is that phenomenology only weakly constrains the study of social cognition, phenomenology would not have any special methodological role in debates about social cognition, and there would be no serious disagreement between mindreading theories and EC accounts of folk psychology.

Highlighting this point of agreement brings into focus the true disagreement between EC theorists and mindreading theorists. The real disagreement concerns whether phenomenology *strongly* constrains the study of social cognition. To settle this disagreement, we have to determine whether phenomenology plays a more substantial methodological role. For instance, must a theory of social cognition explain and be empirically supported by our phenomenological experience? If a theory does not do this, is this grounds for decisively rejecting the theory? Some theorists employ phenomenological evidence in just this way (Gallagher 2012; Zahavi 2011). This is a much stronger constraint that many theories of mindreading would not satisfy.

Whether this stronger constraint on theories of folk psychology is legitimate depends on the nature of phenomenological evidence. Empirical evidence capable of playing a substantial methodological role in theory choice typically is *novel*, *reliable*, and *relevant* to the theories' predictions. Evidence is novel when it provides information beyond what is widely acknowledged and presupposed by the theories debated. It is reliable insofar as it is consistent, accurate, and unbiased. Evidence is relevant to the extent that it can confirm or disconfirm some of the hypotheses debated. In general, empirical evidence that plays a substantial methodological role in theory choice has these three characteristics.[7] Phenomenology, I argue below, has none of these characteristics.

First, phenomenological evidence is not novel in the relevant sense because it does not provide evidence beyond what is widely acknowledged and presupposed by the theories debated. Both mindreading and EC theorists are constructed to be compatible with such evidence. Mindreading proponents do not provide detailed accounts of our phenomenology of social interactions, and they do not regard phenomenology as a good guide to the nature of our cognitive systems, but their descriptions of the theories reflect the presupposition of common phenomenological evidence. In particular, their cautionary notes that mechanisms they describe need not and often do not operate consciously and explicitly (Blackburn 1992, 192; Davies and Stone 1995, 2; Goldman 2006, 12) imply that their theories presuppose that our ordinary social interactions often seem to involve no effortful deliberation about mental states. Thus, pointing out that our phenomenological experiences of social interactions typically are fluid and seem to involve no conscious deliberation does little to mediate the dispute between mindreading and EC theorists.

Second, phenomenological evidence is not reliable. Reliability is operationalized as consistency, accuracy, and objectivity. I go into much more detail about how to understand these dimensions of reliability in Spaulding (2015), but here I will provide just a quick description of how to understand these characteristics. Phenomenological reports are consistent to the extent that they are similar for similar interactions. A subject's reported phenomenology is accurate to the extent that it provides a factually correct description of the experience. And a subject's phenomenological report is objective to the extent that it is not significantly biased by evidentially irrelevant factors.

Many of us are aware of the extensive empirical literature detailing how phenomenological reports vary both across individuals and even with the same individual in different contexts. Our phenomenological reports are influenced by our motivations, cognitive load, situational context, and a host of other factors that influence our attention. Schwitzgebel's (2008) comprehensive review of the empirical evidence details the many ways in which our phenomenological reports are both interpersonally and intrapersonally inconsistent.

Inconsistency would not be problematic if we had standards for what counts as accurate phenomenological introspection. We could just pay attention to the accurate phenomenological reports and dismiss the inconsistent, inaccurate reports. However, in many cases we lack accuracy standards for phenomenological reports. But even when we have some way to judge accuracy of a phenomenological report, phenomenology's track record is not good (Schwitzgebel 2008). When we are distracted, passionate, inattentive, self-deceived, pathologically deluded, reflecting on minor matters, reflecting on the past, reflecting on the current moment, and where fine discrimination is required, we are prone to error. Given the variety of circumstances that lead to error, we can safely conclude that phenomenology often is inaccurate.

Finally, phenomenological reports are not objective representations of what is going on in our minds. Phenomenological reports always are subject to a response bias. When subjects are instructed to report what they are experiencing, they do not simply read off their experiences. They make choices about what is relevant or important and what is not worth mentioning. Without standards of accuracy, it is difficult to see how to correct for response bias. Because phenomenology often is inconsistent, inaccurate, and not objective, we have good reason to doubt the reliability of phenomenological evidence.

The third and final characteristic of evidence capable of playing a substantial methodological role in theory choice is that it is relevant to the theories debated. Phenomenological evidence is *not* relevant to debates about mindreading. The debate about whether mindreading theories make predictions that are testable phenomenologically has gone through several cycles of refinement (Gallagher 2004, 2012; Spaulding 2010, 2015; Zahavi 2011). After a fair bit of talking past one another, it seems that the debate has come down to whether explanation and prediction are exclusively personal-level phenomena and thus whether lacking phenomenological experience of explanation and prediction indicates that explanation and prediction do not occur.

Gallagher explicitly maintains that explanation and prediction are exclusively personal-level phenomena. "Explanation (or theory) seems to mean (even in our everyday psychology) a process that involves reflective consciousness. The term 'prediction' also seems to me to describe a reflective conscious act . . . 'Explanation' and 'prediction' are personal-level terms" (Gallagher 2005, 215). Mindreading proponents, in contrast, use the terms "explanation" and "prediction" in a broader way to refer to both personal and sub-personal processes that are not consciously accessible (Jacob 2011; Spaulding 2010; Herschbach 2008). According to this view, many of our social interactions consist in *tacit* or *implicit* mindreading, i.e., subconsciously explaining and predicting targets' behavior on the basis of attributed mental states.

The terms "explanation" and "prediction" do connote effortful, conscious processing, so it is understandable that Gallagher and others would interpret them as exclusively personal-level phenomena. These terms are carryovers from the early days of the theory of mind literature when philosophers conceived of theory of mind as a literal theory with a Deductive-Nomological structure. No contemporary theorist takes that idea seriously anymore, but the terminology has stuck around. Besides this terminological confusion, it seems like a harmless feature of the literature. If one is keen to avoid the terminological confusion at the heart of this debate about whether explanation and prediction always are effortful, conscious processes, one could substitute "interpretation" and "anticipation" for explanation and prediction. There is no reason to worry that interpretation and anticipation are exclusively personal-level phenomena, and the terms capture what mindreading theorists intend to claim. All of this is to say that the idea that phenomenology is relevant rests on an unfortunate terminological confusion that we can avoid.

EC's phenomenological arguments against mindreading fail because phenomenological evidence is not consistent, reliable, or relevant. Consequently, phenomenology should not play a substantial role in debates about social cognition. Theories of mindreading need not explain and be empirically supported by phenomenological reports about our social interactions. This of course does not settle the entire debate between mindreading theories and EC. The arguments advanced in this section simply put to rest a prominent, common, but misguided objection to the broad scope of mindreading claim. The debate about mindreading is an instance of the more general debate between EC and cognitivism, and that debate centers on mental content and representation (Shapiro 2010; Clark 2008; Hutto and Myin 2012). This more general debate is worth having, but to have those debates at the appropriate level of detail would require at least another set of papers (which I welcome others to write).[8] For now, I will move on to other challenges to mindreading.

2.3 The pluralistic folk psychology challenge

A different kind of challenge to the broad scope of mindreading claim comes from proponents of pluralistic folk psychology, a view that maintains that we have lots of different methods for understanding and interacting with others and mindreading is just one of these methods (Andrews 2012; McGeer 2007). On the pluralistic view, the folk psychological literature has paid far too much attention to mindreading, explanation, and prediction and far too little attention to trait attributions, stereotypes, behavioral schemata, and the way in which we employ these folk psychological tools to regulate our own and others' behavior.

Kristin Andrews (2012), one of the first to propose this view, argues that some non-human animals clearly are capable of attributing personality traits to others but there is little evidence that they are capable attributing propositional attitudes like beliefs. Similarly, people on the autism spectrum are capable of attributing traits and employing schemata for social interactions but nevertheless have difficulty attributing beliefs and desires. These dissociations suggest the attribution of traits, schemata, etc. do not require mindreading. Andrews argues that humans and others species can function socially quite well with these other social practices. She does not deny that adult humans mindread. She argues we mindread to explain anomalous behavior that we are interested in explaining. However, on her view, mindreading is just one amongst many social practices that make up our folk psychological toolkit. Andrews' view is in some ways sympathetic to the EC challenge to the broad scope claim insofar as she agrees that mindreading is not the main tool for social interaction. However, her view differs from EC in that her arguments do not stem from a rejection of cognitivism (and thus, her skepticism is not grounded in a rejection of the idea that our cognition trades in mental representations). Moreover, her arguments are based on empirical evidence of dissociation between mindreading and other social practices rather than an appeal to phenomenology.

Victoria McGeer (2007) offers a related version of folk psychological pluralism that she calls *regulative folk psychology*. On McGeer's view, mindreading is more central to folk psychology than on Andrews' view. McGeer argues that we regularly attribute to ourselves and to others beliefs, emotions, traits, etc. In contrast with the standard mindreading story, however, she emphasizes more than just belief and desire attributions, and she argues that the function of such attributions often is to regulate our own and others' behavior, not to explain and predict behavior. When I attribute to you the trait of being nurturing, I am using this personality trait to make sense of your past behavior but also, and often most importantly, I am implicitly saying to myself, to someone else, or to you that you ought to behave in a nurturing way because that is who you are. Thus, our folk psychological practices often are normative on McGeer's view. The normative role of folk psychology is not at all well captured by the standard mindreading story.

Finally, Tad Zawidzki (2013) presents a view quite similar to McGeer's regulative folk psychology. Zawidzki argues that our folk psychological practices primarily aim at what he calls *mindshaping* rather than mindreading. In ontogenetic development and in adults' ordinary social interactions, our social practices aim to *shape* minds to conform to predictable patterns of behavior. Mindshaping comes in the form of imitation, pedagogy, conformity to norms, attribution of traits, narrative self-constitution, etc. By attributing traits and mental states to others, we are attempting to make others' minds more coherent and more predictable, which enables cooperation, the development of tools and language, and even mindreading. Thus, the primary function of our social practices is to get ourselves and others to conform to certain ways of thinking and behaving to facilitate cooperative engagement. Like Andrews and McGeer, Zawidzki's argument against the broad scope claim is not based on EC-like commitments, and he does not doubt that we engage in mindreading. Rather, like the other views under the pluralistic folk psychology banner, the mindshaping hypothesis maintains that we have many social tools other than mindreading, and the primary function of these tools often is not explanation and prediction.

Two themes of the pluralistic views described above are worth highlighting. First, the views hold that mindreading is just one amongst many social practices. Second, mental state ascriptions are not causal explanations but rather justificatory or rationalizing explanations that regulate our social interactions. These themes do not conflict with the letter of standard mindreading story but are at odds with the spirit of the view. As I mentioned above, mindreading theories are not committed to the claim that we always, in every circumstance, must engage in mindreading to understand others. Not every instance of socializing is underpinned by mental state attribution and explanation and prediction. The difference between the standard mindreading story and pluralistic folk psychology is a matter of emphasis and degree. Pluralistic folk psychology emphasizes the other non-mindreading social practices and holds that mindreading, as traditionally understood, is a distinct and relatively rarely used social practice. The standard mindreading story emphasizes mindreading, unsurprisingly, and neglects to say much of anything about other modes of social interaction.

Pluralistic folk psychology raises two important questions for mindreading theories. First, if one accepts that we can navigate the social world using methods other than mindreading, are mental state ascriptions a necessary or central part of our social interactions? Do we need mindreading at all, or can we just get by with schemata, scripts, personality trait inferences, and stereotypes? The most sensible answer to this question, and one that I will defend in subsequent chapters, is that we have a broad repertoire of methods for understanding others and these methods interact in messy ways. Some of these methods have received little attention from mindreading theorists. For example, as I shall argue in the next chapter, social categorization and stereotypes are deeply influential in our social interactions. They shape how we see others' behaviors before we even attempt to interpret their mental states. The influence of social categorization and stereotypes is so deep that we might not even recognize that we are employing these social practices until someone with different ways of categorizing and different stereotypes points out that they see the social interaction radically differently. Social categorization and stereotypes have received very little attention in the mindreading literature, which is puzzling and unfortunate because they strongly influence the mental state attributions we end up making when we do engage in mindreading. Thus, an adequate account of mindreading ought to take into consideration the effect that these other social practices have on mindreading (see, for example, Spaulding 2017 and Westra 2017a, 2017b).

The influence goes the other way around, too. Mental state ascriptions can influence the personality trait inferences we make and the scripts we employ. For example, inferring that someone is nurturing or ambitious may require that you make assumptions about their desires and intentions, which may make certain schemata or scripts more readily available. In real-life social interactions (as opposed to abstract, simplified examples employed in the philosophical and empirical literature), these various social practices interact and influence each other in messy ways. We are never *just* attributing a personality trait or *just* attributing a desire or *just* attributing a stereotype. We often are doing all of these things without much reflection on the subtle and complex ways in which mental state attributions, trait attributions, schemas, and stereotypes influence each other. Thus, there is good reason to doubt we can pull apart these social practices in real-life social interactions. This puts pressure on the idea sometimes implicit in pluralistic folk psychology that we can separate mindreading from these other social practices (Andrews 2012; Zawidzki 2013). We cannot, which is all the more reason for the mindreading theorists to embrace the spirit of the challenge from pluralistic folk psychology. In Chapter 3, I shall argue that we ought to expand mindreading theories to include these other social practices.

A second question that pluralistic folk psychology raises is whether mental state ascriptions serve as causal explanations and predictions or rationalizing and regulating explanations. The most sensible answer seems to be that our mental state ascriptions serve both of these roles, and others as well. Again, it is messy. As I will argue in Chapter 4, theorists in the folk psychological literature (broadly construed to include mindreading, EC, and pluralistic folk psychology) have paid virtually no attention to the variety of goals we have in a social interaction, the different approaches to mindreading that correspond to these goals, and the different kinds of products of mindreading. Our goals and corresponding approaches to mindreading have a profound impact on the kind of explanation we generate and what we do with that explanation. I will discuss this in much more detail in Chapter 4. The point I want to make here is that mindreading serves many purposes. In real-life social interactions, the different explanatory roles of mindreading are difficult to pull apart because they shift as our goals and the dynamics of an interaction change. It is not helpful to assert a dichotomy between causal explanations and rationalizing explanations because mindreading can generate many different kinds of products. I shall argue in Chapter 4, we have good reason to support pluralism about our repertoire of social practices, the kinds of explanation we generate in social interactions, *and* what we do with those explanations. Indeed, a broad, overarching theme of this book is that our social interactions are much messier, more complicated, and more diverse than the empirical and philosophical literatures on folk psychology indicate.

2.4 Assessment

I have considered several challenges to the idea that mindreading is a central and common social practice. I argued that EC's main challenge to the broad scope claim – the phenomenological argument – fails because phenomenology cannot play a substantial methodological in these debates. The evidence it provides is not novel in the relevant sense because mindreading theories are constructed to be compatible with such phenomenological evidence. Phenomenological introspection is infamously unreliable. And the veneer of relevance disappears once we realize the unfortunate terminological confusion at the heart of the debate. Thus, I do not find much compelling in the EC challenge to the broad scope of mindreading claim.

In contrast, the challenge from pluralistic folk psychology puts pressure on the broad scope claim in compelling ways. Pluralistic folk psychology advocates for attention to social practices other than mindreading and roles other than explanation and prediction for these social practices. Careful attention to real-life social interaction shows that mindreading is deeply intertwined with all sorts of social practices and types of explanatory inferences. Because of the complicated, interconnected nature of all these social practices, I regard the challenge from pluralistic folk psychology as an opportunity rather than an objection to the standard mindreading account. The pluralistic account does not show that mindreading is less important and rarer than the mainstream mindreading literature maintains. It shows that our social interactions are complicated, and if an account of mindreading is going to be descriptively accurate it ought to capture the diversity and nuance of our social interactions. Chapters 3 and 4 focus on detailing the various cognitive processes at play in social interaction, and Chapter 5 develops a more comprehensive and empirically adequate account of mindreading.

Notes

1 This chapter draws on previously published work on embodied and enactive cognition and the role of phenomenology in debates about social cognition. See Spaulding (2010, 2015) for more in-depth analysis of these topics.

2 There are extensive, ongoing, difficult debates about Autism Spectrum Disorder and theory of mind (Adams 2013), when and how children come to have mature mindreading abilities (Carruthers 2013), and the extent to which non-human animals possess theory of mind abilities (Andrews 2012). For the most part, I will sidestep these thorny discussions in this book.

3 Several features of this typical example of mindreading are worth highlighting. First, the example focuses only on beliefs and desires and not any other cognitive states. Second, the setup of the example suggests that we are often aiming to accurately explain others' mental states. Finally, the example does not specify the kind of explanation at play, but it seems to be a teleological explanation. Each of these features of typical mindreading cases will come up again when I develop my own critiques of the standard mindreading story later in this chapter and in subsequent chapters.

4 De Jaegher and Di Paolo (2007) and De Jaegher, Di Paolo, and Gallagher (2010) offer a more radical alternative to the orthodox view of mindreading. On this interactionist proposal, social interaction *constitutes* social cognition. This view has the advantage of being able to explain a lot of the effects of situational context I describe in later chapters, but in my view, interactionism is weakly motivated and stumbles on issues of constitution and ontology.

5 I shall talk a lot more about this in Chapters 3 and 4.

6 The ensuing arguments regarding the weak and strong methodological roles of phenomenology draw heavily from Spaulding (2015). That paper goes far beyond the discussion here. For those interested in a more detailed discussion of the philosophy of science and empirical evidence that I outline here, see Spaulding (2015).

7 I do not discuss how we ought to order and weight these features or if in every case of theory choice all three features are necessary. In this case, these details are irrelevant because I will argue that phenomenological evidence about social interactions has none of these features.

8 Part of the concern about EC as an alternative to cognitivism is that many of the arguments simply critique cognitivist proposals rather than construct a coherent alternative to cognitivism, and the arguments against mental content and representations often are *very* flimsy, even according to people who are generally sympathetic to the EC research program (Shapiro 2014).

References

Adams, M. 2013. "Explaining the theory of mind deficit in autism spectrum disorder." *Philosophical Studies* 163 (1): 233–249.

Andrews, K. 2012. *Do Apes Read Minds? Toward a New Folk Psychology*. Cambridge, MA: MIT Press.

Baron-Cohen, S. 1995. *Mindblindness: An Essay on Autism and Theory of Mind.* Cambridge, MA: MIT Press.

Blackburn, S. 1992. "Theory, observation and drama." *Mind & Language* 7 (1–2): 187–230.

Carruthers, P. 2009. "How we know our own minds: The relationship between mindreading and metacognition." *Behavioral and Brain Sciences* 32 (2): 1–18.
Carruthers, P. 2013. "Mindreading in infancy." *Mind & Language* 28 (2): 141–172.
Clark, A. 2008. *Supersizing the Mind: Embodiment, Action, and Cognitive Extension (Philosophy of Mind Series)*. New York: Oxford University Press.
Davies, M., and T. Stone. 1995. *Folk Psychology: The Theory of Mind Debate*. Oxford: Blackwell.
De Jaegher, H., and E. Di Paolo. 2007. "Participatory sense-making." *Phenomenology and the Cognitive Sciences* 6 (4): 485–507.
De Jaegher, H., E. Di Paolo, and S. Gallagher. 2010. "Can social interaction constitute social cognition?" *Trends in Cognitive Sciences* 14 (10): 441–447.
Fodor, J. A. 1987. *Psychosemantics*. Cambridge, MA: MIT Press.
Gallagher, S. 2001. "The practice of mind: Theory, simulation or primary interaction?" *Journal of Consciousness Studies* 8 (5–7): 83–108.
Gallagher, S. 2004. "Understanding interpersonal problems in autism: Interaction theory as an alternative to theory of mind." *Philosophy, Psychiatry, & Psychology* 11 (3): 199–217.
Gallagher, S. 2005. *How the Body Shapes the Mind*. New York: Oxford University Press.
Gallagher, S. 2012. "In defense of phenomenological approaches to social cognition: Interacting with the critics." *Review of Philosophy and Psychology* 3 (2): 187–212.
Gallagher, S. 2017. *Enactivist Interventions: Rethinking the Mind*. Oxford: Oxford University Press.
Goldman, A. I. 2006. *Simulating Minds: The Philosophy, Psychology, and Neuroscience of Mindreading (Philosophy of Mind)*. New York: Oxford University Press.
Herschbach, M. 2008. "False-belief understanding and the phenomenological critics of folk psychology." *Journal of Consciousness Studies* 15: 33–56.
Hobson, P. 2004. *The Cradle of Thought: Exploring the Origins of Thinking*. London: Pan Macmillan.
Hutto, D. D., and E. Myin. 2012. *Radicalizing Enactivism: Basic Minds Without Content*. Cambridge, MA: MIT Press.
Jacob, P. 2011. "The direct-perception model of empathy: A critique." *Review of Philosophy and Psychology* 2 (3): 519–540.
Leslie, A. M. 2000. "Theory of mind as a mechanism of selective attention." *The New Cognitive Neurosciences*: 1235–1247.
McGeer, V. 2007. "The regulative dimension of folk psychology." In *Folk Psychology Re-assessed*, 137–156. New York: Springer.
Schwitzgebel, E. 2008. "The unreliability of naive introspection." *Philosophical Review* 117 (2): 245–273.
Shapiro, L. A. 2010. *Embodied Cognition*. New York: Routledge.
Shapiro, L. A. 2014. *Radicalizing Enactivism: Basic Minds Without Content*, edited by Daniel D. Hutto and Erik Myin. Oxford: Oxford University Press.
Spaulding, S. 2010. "Embodied cognition and mindreading." *Mind & Language* 25 (1): 119–140.
Spaulding, S. 2011. "A critique of embodied simulation." *Review of Philosophy and Psychology* 2 (3): 579–599.

Spaulding, S. 2015. "Phenomenology of social cognition." *Erkenntnis* 80 (5): 1069–1089. doi:10.1007/s10670-014-9698-6.

Spaulding, S. 2017. "Do you see what I see? How social differences influence mindreading." *Synthese*.

Tooby, J., and L. Cosmides. 1995. "Foreword to Baron-Cohen." In *Mindblindness: An Essay on Autism and Theory of Mind*. Boston: MIT Press.

Westra, E. 2017a. "Character and theory of mind: An integrative approach." *Philosophical Studies*.

Westra, E. 2017b. "Stereotypes, theory of mind, and the action-prediction hierarchy." *Synthese*: 1–26.

Zahavi, D. 2011. "Empathy and direct social perception: A phenomenological proposal." *Review of Philosophy and Psychology* 2 (3): 541–558.

Zawidzki, T. W. 2013. *Mindshaping: A New Framework for Understanding Human Social Cognition*. Cambridge, MA: MIT Press.

3 Expanding the scope of mindreading

Social categorization, stereotypes, social bias, and situational context

In this chapter I argue that existing mindreading theories often are too narrowly focused on belief attribution. As a result, they neglect other aspects of social interaction, such as how automatic social categorization and situational context influence how we interpret social behavior. These other aspects filter input to mindreading and thus directly influence the mindreader's mental state attributions. I explain how social categorization and situational context prime certain interpretative schemata, activate stereotypes and social biases, and make certain personality trait and mental state inferences salient and sometimes hard to override.[1]

3.1 The narrow focus of standard mindreading theories

Though I gave a brief overview of some general theories of mindreading in the introductory chapter, to see how the mindreading literature came to have such a narrow focus it is important to recall how the empirical and philosophical study of mindreading began. The discussion of mindreading, also called theory of mind, began in earnest with David Premack and Guy Woodruff's (1978) study, which examined whether chimpanzees attribute beliefs to other chimpanzees. This foundational study ignited much philosophical and empirical work on belief attribution. On the basis of Premack and Woodruff's study, Daniel Dennett (1978) and others theorized about what kind of evidence would demonstrate that an agent possesses the concept of belief and could therefore attribute a belief to others. The problem with theory of mind tests up to this point was that they could not distinguish merely apparent belief attribution from genuine belief attribution. Dennett's rationale was that to distinguish whether an agent really can attribute beliefs, we have to test whether she can attribute *false* beliefs. If an agent can attribute a false belief, then she surely grasps that beliefs represent the world a certain way and that some agents' beliefs can *misrepresent* the world.

Dennett's work inspired Hans Wimmer and Josef Perner (1983) to develop what is now known as the standard false-belief task. In one version of this task, subjects watch a puppet put a toy in one location and leave the scene. Next, they watch another puppet come in and move the toy and then leave the scene. Finally, they watch the first puppet return to the scene, and they are asked where the first puppet thinks her toy is. Wimmer and Perner made the revolutionary discovery that before age 4 virtually all children systematically fail the standard false-belief test, but after age 4 virtually all children pass this test. This result inspired decades of research on who, when, how, and why we attribute belief. Even the critiques of the standard false-belief task (Bloom and German 2000) inspired more work on belief comprehension – non-linguistic versions of the false-belief task (Baillargeon, Scott, and He 2010; Onishi and Baillargeon 2005) and vigorous debate about whether these non-linguistic false-belief tasks really are tracking beliefs (Heyes 2014; Apperly and Butterfill 2009) – rather than a shift away from belief attribution. The implicit rationale of this focus on belief seems to be that belief attribution is a difficult and fundamentally important aspect of theory of mind. Once one grasps belief attribution, the rest of theory of mind will fall in place, and so the research should focus on issues surrounding belief attribution.

The debate between the TT and the ST grew out of this discussion about belief attribution. In the early days of this literature, it was widely accepted that the TT and the ST were mutually exclusive general accounts of mindreading. These days, no one subscribes to *pure* TT or *pure* ST. The problems associated with monolithic versions of the TT and the ST are well documented (Davies and Stone 1995a, b), and both the TT and the ST can boast empirical vindications. Most contemporary mindreading theorists hold some kind of hybrid account. Hybrid accounts aim to capture the theoretical advantages of ST and TT while avoiding the problems with both theories. As I describe in the introductory chapter, contemporary versions of the TT and the ST are much more sophisticated and nuanced than the initial versions of the TT and the ST. What remains the same, however, is the focus on who, when, how, and why we attribute beliefs to others.

Even various subfields of mindreading take belief attribution to be a central *explanandum*. Debates about mindreading and introspection concern how we attribute beliefs to ourselves (Carruthers 2011). Debates about mindreading and Autism Spectrum Disorder concern why people with autism have difficulty attributing beliefs (Oberman and Ramachandran 2007). Debates about mindreading and pretense are about whether engaging in and recognizing pretense requires belief attribution (Nichols and Stich 2000). Debates about the ontogenetic development of mindreading concern when and how children learn to attribute beliefs to others (Baillargeon, Scott, and He 2010; Heyes 2014). Mindreading discussions, both general and specific, are framed by questions about how a mindreader attributes beliefs to a target. In all of these debates, theorists aim to explain how we understand and interact with others by explaining how a subject comes to attribute beliefs (and to some extent other mental states) to a target.

This fixation on belief attribution is unfortunate, because this is only one tiny part of the story of how we understand and interact with others. Left out of the standard mindreading story are the ways in which situational context influences in-group dynamics, which behavioral patterns and stereotypes are salient, the personality trait inferences we make in various contexts, and the biases that shape how we interpret a social interaction.[2] These factors play a role in filtering the available information in a subject's environment, what *kind* of mindreading we engage in, and the kind of mindreading product we generate. My claim is not simply that these factors influence how we subsequently use our mindreading judgments. Rather, I shall argue in this chapter and the next that the input, processing, and output of mindreading vary along many dimensions, which makes constructing an empirically adequate account of mindreading significantly more challenging than typically recognized.

3.2 Social categorization

Social categorization consists in sorting people, behaviors, and events into social categories. Such sorting is essential for successful navigation of the social world. It helps make the social world more comprehensible and predictable, and thereby allows us to manipulate the social world for our purposes. Without the ability to detect patterns and categorize people, situations, and events, we would be incapable of fluidly navigating the social world.[3] The fact that we divide people up in particular ways does not imply that these divisions are genetically or metaphysically robust. It may be the case that some categories we employ are merely socially constructed and there are no real essences of such categories. Nevertheless, we do divide people into social categories.

Social categorization is reflexive and very fast. Within 100 milliseconds of seeing a face, we can sort people by age, gender, and race (Ito, Thompson, and Cacioppo 2004; Liu, Harris, and Kanwisher 2002). We tend to sort people by the most salient category, and age, gender, and race often are the most salient categories in heterogeneous, multicultural societies. However, the categories we employ can be modulated by various factors (Wheeler and Fiske 2005; Gilbert and Hixon 1991). Context, cognitive load, and our goals influence what is salient to us, and what is salient determines which social categories we employ when we reflexively sort people into social groups. For example, in a competitive context where team affiliation is salient (e.g., at a sporting event), age may be less salient than whether someone is wearing certain colors. In a context where nationality is salient (e.g., in an airport's international terminal), race and gender may be less salient than someone's accent. Though we tend to sort people by age, race, and gender most frequently and most rapidly, classifications like nationality, religion, regional geographical identity, and class also are commonly employed social categories.

On the basis of rapid facial recognition, we spontaneously infer personality traits, as well, e.g., trustworthiness, competence, aggressiveness, or dominance (Olivola and Todorov 2010; Rule, Ambady, and Adams Jr 2009). We make such inferences very quickly (somewhere between a tenth of a second to a second and a half) even when we are under cognitive load (Todorov and Uleman 2003; Malle and Holbrook 2012). These inferences on the basis of very narrow slices of behavior surprisingly turn out to be fairly accurate (Palomares and Young 2017). Inferences about personality traits play an important role in the mental states we attribute to people and our interpretation of their behavior. If I infer that someone is, say, aggressive (on the basis of a very narrow slice of behavior, like a glimpse at his facial expression) this will influence how I interpret his behavior, whether I perceive him to be like me or different, whether I like him or not, etc. All of these subtle effects of spontaneous personality trait inferences shape my perspective on the target before I even begin to form beliefs about his mental states (Westra 2017a).

3.3 Stereotypes

Social categories are associated with various features, and these associations form the basis of social stereotypes. Stereotypes are conceptually rich systems of belief about social groups (Liberman, Woodward, and Kinzler 2017). As I use the term, stereotypes can be positive, negative, or neutral beliefs about members of particular social groups.[4] Stereotypes are based on social categorization insofar as stereotyping presupposes a capacity to sort people into social groups, but they are further informed by our learned associations between social groups and particular characteristics.[5]

We employ stereotypes in all sorts of social interactions. We rely on beliefs about how certain groups of people behave, and this helps us interpret their behavior quickly and efficiently, and it helps us anticipate what they will do next (Westra 2017b). Though there are many real-life examples of stereotype use, fiction provides a vivid illustration of how ubiquitous stereotypes are.

Consider as an example the strangely popular television sitcom *The Big Bang Theory*. The characters in the show represent particular stereotypes. Sheldon and his science friends represent a common stereotype of academic scientists: relatively unattractive, nerdy, bad at communicating about topics other than science, and romantically challenged. Penny, the aspiring-actress waitress down the hall, represents a contrasting stereotype: a ditzy, unintelligent, yet charming and attractive young female. Understanding the interactions of the characters in this show requires knowledge of the relevant stereotypes. If one did not grasp the stereotypes that Sheldon and Penny represent, one would be baffled by their odd interactions. However, because the audience is adept at applying these stereotypes, these interactions are completely understandable (even utterly, painfully predictable to some of us).

Works of fiction, like novels, plays, television shows, movies, and comic books, implicitly rely on the audience's grasp of the relevant stereotypes. Without background knowledge of the stereotypes, we would fail to make sense of characters' behaviors and interactions. Of course, *good* fiction portrays characters that go beyond the stereotypes they represent. Characters who exhibit only stereotypical behavior are completely predictable, and we tend to find such flat characters dull and uninteresting. However, many fictional characters are at least *based on* well-known stereotypes. A fictional character is interesting insofar as she or he departs from the stereotype. We find such departures surprising, amusing, and at times horrifying. Importantly, however, we would not have such reactions if we did not understand that the behavior was a departure from a stereotype. We have these reactions precisely because we recognize the stereotype and form expectations based on that stereotype and are surprised when the characters violate those expectations. Thus, in consuming fiction, the audience deftly applies stereotypes as a way of making sense of characters' behaviors, even those behaviors that deviate from the stereotype.

The same is true in real-life interactions. We often employ stereotypes in interpreting and anticipating people's behavior. Stereotypes help us rapidly form expectations for how certain individuals will behave. Even when we interpret or anticipate behavior that deviates from the stereotype, it is often against the backdrop of a relevant stereotype. This is an important influence on mindreading that is relatively unacknowledged in the mindreading literature. In this section I have presented stereotypes as an ethically neutral but significant influence on how we interpret social interactions. In the next section, I shall discuss how stereotypes and other social biases get put to use in more nefarious ways.

3.4 Social biases

As I mention in the previous section, the categories we use to sort people are associated both explicitly and implicitly with various characteristics.[6] Explicit associations are relatively easy to test because subjects have no difficulty reflecting on and reporting their explicit associations. You can simply ask people what characteristics they associate with various social categories and try to control for social desirability censorship. Implicit associations are more difficult to test because subjects often are unable or unwilling to report their implicit associations. For this reason, experimenters construct tasks that are designed to elicit behavior that is sensitive to implicit associations, and from the elicited behavior they estimate subjects' implicit associations.

The main experimental paradigms for investigating implicit associations are the Implicit Association Test (IAT) and various types of priming tasks. The IAT measures how quickly and accurately subjects categorize stereotypic and counter-stereotypic associations (Greenwald, McGhee, and Schwartz 1998; Greenwald et al. 2009). In one well known version of the IAT, subjects are instructed to categorize as quickly and accurately as possible faces of Black men with pleasant words (e.g., "joy," "love," "peace") and faces of White men with unpleasant words (e.g., "agony," "terrible," "horrible"). Subjects are then instructed to categorize the stimuli according to the opposite rule: Black faces with unpleasant words and White faces with pleasant words. If subjects categorize faster and more accurately according to one of these rules, they are said to have an implicit bias. As it turns out, most White Americans more strongly associate Black with unpleasant words and White with pleasant words than Black with pleasant and White with unpleasant. This is taken as evidence that most White Americans are implicitly biased against Black people.

Priming tasks are a second method for investigating implicit associations. Priming tasks measure the effects of subtle cues in the environment on our emotional and cognitive responses (DeCoster and Claypool 2004; Fiske and Taylor 2013, 3–37). Subliminal priming occurs when a stimulus is presented to subjects too quickly to be consciously processed. Conscious priming occurs when the subject consciously perceives the prime but has no awareness of its effects on subsequent reactions. Priming tasks can be engineered to uncover cognitive priming, in which subtle cues in the environment affect subjects' judgments, and affective priming, in which subtle cues in the environment affect subjects' emotional states.

There are many different types of priming tasks. The following study is one illustration of how a cognitive priming task measures implicit associations. Graham and Lowery (2004) report that police officers and juvenile probation officers subliminally primed with words related to the racial category Black are more likely to interpret a hypothetical adolescent (whose race is unspecified) as having a worse personality, being more blameworthy, and being more likely to reoffend, and they recommended harsher punishments. This association between the racial category Black and delinquency is the sort of thing that subjects may be unable or unwilling to articulate to experimenters, but nevertheless it is salient in their minds and affects their judgments. For our purposes here, the relevant point is that the associations may influence which features of a situation subjects notice, which shapes the information base for the mindreading judgments subjects go on to make.

In philosophy and the cognitive sciences, implicit bias and the tests used to measure implicit bias have become quite controversial. The intensity of the debates about implicit bias warrants a short detour to explain the controversy. Skeptics about implicit bias argue that there is no single mechanism that could explain the various behaviors attributed to implicit bias, some of the most prominent measures of implicit bias have poor test-retest reliability, and measures of implicit bias sometimes predict biased behavior but other times do not.[7] A modest skeptical claim is that theorizing about implicit bias has outpaced empirical measurement of implicit bias, and the focus should be on developing better empirical measures rather than applying the implicit bias construct to myriad theoretical and practical issues. A related worry is that the overwhelming focus on *implicit* bias overshadows the work on *explicit* bias, which is very real and very problematic. The most radical skeptical claim is that there is no such thing as implicit bias, and the whole discussion is wrongheaded.[8]

Because the skeptical view, if right, could derail this part of my project, and because this is an important topic in its own right, I will explain my own view on these skeptical worries. My view is that the skeptics' objections are legitimate concerns, but they do not warrant skepticism about the phenomenon of implicit bias. In Del Pinal and Spaulding (2018), my co-author and I develop a model of bias according to which social biases can be encoded in different cognitive structures. On this view, a social bias just is an association between a social category and a feature. Consider the association between the category *woman* and the feature *family oriented*. One's association between women and family orientation may reflect salience, e.g., family orientation is prominent in one's conception of women, or the inference that a woman is family oriented is readily available. Alternatively, one's association may reflect statistical information, e.g., women represent a higher percentage of all the people who are family oriented, or most women are family oriented. See Table 3.1 below.[9] IATs and priming measures are constructed to detect these salient-statistical associations by eliciting behavior that is sensitive to context. Priming tasks investigate subtle effects of stimuli on cognitive and affective processes, but when these stimuli are altered or presented in different contexts, the results differ (Wittenbrink, Judd, and Park 2001). Similarly for IATs, when the stimuli include context or richer representations, e.g., well known figures rather than generic faces, the results differ (Govan and Williams 2004). This is evidence that priming

Table 3.1 Salient-Statistical Associations

Relation/term	*Types of measures*	*Examples*
Saliency: Signal-to-noise ratio	Prominence	How prominent in your conception of women is it that they are family oriented?
	Availability	Are women family oriented? (yes/no)
Statistical: Evidence provided by a feature for a category	Cue validity: P(category|feature)	Of all people who are family oriented, what percentage are women?
	Typicality: P(feature|category)	Of all the women, what percentage are family oriented?

tasks and IATs are successfully tracking highly context-sensitive salient-statistical associations.

Salient-statistical associations are not the only way in which biases are encoded in cognitive structures. Salient-statistical associations systematically dissociate from more stable biases that are encoded in dependency networks. A dependency network represents the relations amongst various features associated with a concept. Some features are more peripheral insofar as one can easily imagine an instance of that concept without that feature (e.g., a mane-less lion) and some features are more central insofar as one cannot easily imagine an instance of that concept without that feature (e.g., a lion that is not a mammal). Importantly, salient features need not be central features (e.g., a mane is a salient feature of lions, but it is not a central feature).

Biases encoded in dependency networks reflect how central a feature is in one's conception of that social category. See Table 3.2 below.[10] For example, a centrally encoded bias that women are family oriented would be reflected in a subject's surprise at encountering a woman who is not family oriented (compared to the surprise at encountering a man who is not family oriented), or the ease of imagining a woman who is not family oriented compared to the ease of imagining a man who is not family oriented. This framework for understanding bias posits various distinct cognitive processes underlying implicit bias, which systematically dissociate from each other. Centrally encoded biases are much more resilient across contexts than salient-statistical associations, and as a result they behave quite differently in experimental contexts. The focus of the philosophical and empirical literature overwhelmingly has been on salient-statistical associations, but these are the most unstable, context-sensitive biases. Centrally encoded biases display much more cross-contextual stability and as a result can be much more ethically worrying.

Table 3.2 Conceptually Central Associations

Relation/terms	*Types of measures*	*Examples*
Conceptual centrality	Surprise	How surprised would you be to encounter a woman who is not family oriented?
	Ease-of-imagining	How easily can you imagine a real woman who is not family oriented?
	Goodness-of-example	How good an example of a typical woman would you consider a woman who is not family oriented?
	Similarity-to-an-ideal	How similar is a woman who is not family oriented to an ideal woman?

The benefits of this model of bias are that it applies to both implicit bias and explicit bias, it makes correct predictions about the context-sensitivity of some forms of implicit bias,[11] it promises to shed light on murky debates about whether implicit biases are beliefs, and it provides an empirical framework for future investigation. This account embraces the skeptics' concerns about implicit bias without accepting the skepticism about the phenomenon of implicit bias. One need not accept this model of bias to accept the argument advanced in this chapter and the book more generally, but it does help to articulate how I understand implicit bias and the roles it plays in social cognition. With that, our brief digression on implicit bias ends.

Switching gears now, another element of social bias that affects mindreading is our habitual tendency to sort individuals into in-groups and out-groups (Tajfel 1974). Sorting people into in-groups and out-groups is a function of perceived similarity, which is a subjective, contextually relative, and sometimes idiosyncratic judgment (Ames 2004a; Ames 2004b). On the basis of the situational context, social categorization, and personality trait inferences, we quickly identify people as "like us" in some relevant respect and therefore part of our in-group or as "unlike us" in some relevant respect and therefore part of an out-group. We may judge someone to be similar to us in terms of racial or ethnic demographics, social class, educational achievement, career, team affiliation, ideals and goals, etc. There are many different ways of sorting people into in-groups and out-groups.

In heterogeneous societies, the default lines of in-grouping/out-grouping are age, race, and gender, because these often are salient features of people in diverse societies. Our social categories extend beyond these classifications, and people have multiple, overlapping identities. I, for example, identify as a teacher, an empirically oriented philosopher, a Wisconsin Badger, a runner, a mother, a wife, a liberal, etc. Who I count as part of my in-group will depend on the context. In a context where career is salient, I regard academics – or perhaps more narrowly, philosophers or even philosophers of cognitive science – as part of my in-group. In a context where family life is salient, I regard parents of young children as part of my in-group, regardless of their careers. In both contexts, I take people in my in-group to share important, relevant similarities. In making such judgments, we tend to gloss over differences within the groups and exaggerate differences between the groups (Linville, Fischer, and Salovey 1989; Mullen and Hu 1989). My academic friends may not be as similar to me and to each other as I assume, and my parent and non-parent friends may not be as different as I assume.

We tend to have more favorable attitudes toward and empathize more with in-group members, especially people who share our gender, race, age, religion, or nationality, than toward people do not share these features (Rudman et al. 1999). In-group favoritism is present even in young children, who prefer own-race faces and own-language speakers (Liberman, Woodward, and Kinzler 2017). Developmentally, in-group favoritism shifts over time, from infants' simple preference for the familiar types of people to elementary-aged children perceiving less humanness in faces of out-group members (McLoughlin, Tipper, and Over 2017).

The effects of in-group favoritism are especially strong in a context of competition or threat, where innocuous preferences for members of one's in-group turn into more pernicious out-group bias (Cikara et al. 2014). This competition or threat could present in an economic guise, such as when members of an economically declining social group feel that they are being asked to sacrifice economic opportunity to help other groups. This is the conceit of J. D. Vance's (2016) memoir *Hillbilly Elegy*, which was published in the run-up to the 2016 U.S. presidential election. Vance argues that over the past several decades White working-class Americans have lost faith in society's ability to provide opportunity for their upward mobility. The economic outlook for this group has declined over the past few decades, whereas other demographic groups have made some economic gains. White working-class Americans, he argues, perceive a threat to their economic well-being. As a result, they are especially sensitive to signals of support for their in-group and signals of opposition toward out-groups who represent a perceived threat or competition. This is one way in which especially strong in-group favoritism can present.

Strong in-group favoritism/out-group bias also occurs when a group perceives a threat to their *values*. Some readers may recall the backlash against then–presidential candidate Barack Obama when he remarked that working-class voters in old industrial towns decimated by job losses "get bitter, they cling to guns or religion or antipathy to people who aren't like them or anti-immigrant sentiment or anti-trade sentiment as a way to explain their frustrations."[12] Conservatives found Obama's comments patronizing and elitist. They circled the wagons and defended their values as legitimate expressions of who they are, not pathetic rationalizations of their situation. More recently, some readers may recall a similar backlash against Democratic presidential candidate Hillary Clinton, who characterized some of Donald Trump's supporters as deplorable. She said "You know, to just be grossly generalistic, you could put half of Trump's supporters into what I call the basket of deplorables. Right? The racist, sexist, homophobic, xenophobic, Islamophobic – you name it." Conservatives pounced on this comment, as they did with Obama's comment, and argued that it was another elitist attack on conservative values.[13] These are two prominent examples of a perceived attack on a group's values intensifying in-group favoritism and corresponding out-group bias.

In contexts of competition and threat, we tend to show strong preferences for our in-group and strong out-group bias. These tendencies manifest in regarding in-group members as more capable of experiencing secondary emotions (such as pride and guilt) and as having richer, more complex mental experiences than out-group members (Haslam 2006). We tend to attribute more simplistic, caricatured mental states to those we perceive to be unlike us. We offer more charitable interpretations of the behavior of those we regard as part of our in-group and tend not to look for or notice similarly charitable interpretations of the behavior of out-group members. Indeed, recent empirical evidence suggests that we are more likely to focus on mental state terms when we are explaining the behavior of in-group as opposed to out-group members (McClung and Reicher 2017). Even 5- to 6-year-old children are more likely to spontaneously attribute mental states to in-group members than out-group members (McLoughlin and Over 2017). These data suggest that we tend to think about the motivations, reasons, and emotions that bring about in-group members' actions, but when it comes to out-groups, we simply consider the behavior itself without thinking about the mental causes of that behavior. All of this indicates that in-group/out-group dynamics have a significant, though subtle effect on how we interpret others' behavior.

A concept related to in-group favoritism is prejudice. Prejudice is based on our capacity to sort people into in-groups and out-groups, but it of course goes much beyond that mere capacity (Liberman, Woodward, and Kinzler 2017). Prejudice further involves negative affect toward out-group members, which may initially begin as in-group favoritism and grow into something more pernicious with increased perceptions of competition or threat. Prejudice also involves negative stereotypes about the target out-group. Thus, an individual who is prejudiced toward some group will feel antipathy toward members of that group and believe negative stereotypes about members of that social group. If one is prejudiced toward particular social groups, one's interpretation of the behavior and mental states of members of that group certainly will differ from the interpretation of someone who is not prejudiced toward that group.

Social biases, such as stereotypes, implicit associations, in-group favoritism, out-group bias, and prejudice, deeply influence how we perceive social interactions. The effects of social bias on mindreading have been almost completely neglected in the mainstream mindreading literature.

3.5 Situational context

In this section, I discuss the influence of the situational context on how we understand and interact with others. Situational context affects every phenomenon described in this chapter. It affects which social categories are salient, the personality trait inferences me make, salient stereotypes, and whether and to what extent social biases are activated. It is thus the most pervasive influence on mindreading, and yet it is rarely discussed in the mindreading literature.

Individuals obviously know that they are in a certain situational context. For example, when I am at an airport, I am aware that I am at an airport and I am aware of the particular norms and behavioral scripts common at airports. However, the influence of the situational context typically is opaque to individuals in that situational context. Usually I do not form a belief that I am at an airport and think about the ways in which airport norms and scripts shape my interpretations and expectations of others' behavior. Rather, I simply act, interpret, and respond to others according to the norms and scripts appropriate for airports. Mindreading theories tend to focus on the process of attributing beliefs. As a result, mindreading theories ignore aspects of the situation that are in the background of mental state attribution. This tendency is to the detriment of mindreading theories because it results in overlooking important phenomena that affect the mental states that mindreaders end up attributing. The situational context is a particularly influential factor in this respect.

The situational context frames our social interactions, e.g., as one of competition, collaboration, or polite indifference. The situational context makes certain features relevant to social categorization salient to us, e.g., others' age, ethnicity, or social class. On the basis of these salient features, we infer personality traits and perceive others as part of a relevant in-group or an out-group, which, as I describe in the previous section, significantly affects how much we like and charitably interpret others, and even whether we attribute mental states at all to someone perceived to be very unlike us. In addition, the situational context shapes our expectations and interpretations of others' behavior. That is, the situational context primes us to expect certain kinds of behavior from members of particular social groups, and these expectations deeply influence how we interpret behavior in that context.

As an example, consider again the juvenile probation officers from the Graham and Lowery study described above. The study indicates that these officers associate the racial category Black with delinquency.[14] On the basis of these findings, we can plausibly speculate that when these officers are in the context of patrolling an urban street corner, they have expectations that the Black youth they see on the street corner are up to no good (especially if they do not personally know the individuals). These expectations, formed on the basis of implicit and explicit associations and richer stereotypes, shape how the officers interpret the behavior they see. The officers' interpretations of the individuals' behavior are shaped, of course, by what the individuals do, but they are also shaped by what they expect the individuals to do. They expect delinquency so they are more likely to notice evidence of delinquency and interpret behavior as consistent with their expectations. So long as the behavior does not overtly thwart expectations, the officers will interpret the individuals' behavior in terms of what they expect to see in that context. Importantly, the very same behavior in a different context (say, these same individuals on a church playground, or White youth in an affluent predominantly White suburban neighborhood) would be interpreted very differently (Gilbert and Hixon 1991; Trope 1986). This pattern of inference is not specific to juvenile probation officers. It is simply the result of one's associations, stereotypes, and confirmation bias, and these are universal cognitive phenomena. In general, these cognitive phenomena make social interactions proceed more smoothly.[15] They become ethically troubling when the associations and stereotypes involved are misguided and pernicious.

The situational context a mindreader finds herself in deeply influences social categorization, in-group/out-group inferences, and the activation of social biases, all of which play an important role in filtering the information that serves as input for mindreading. Focusing exclusively on the front-end product of mindreading – subjects' attribution of a belief to a target – yields an incomplete picture of mindreading. If one neglects factors like the situational context, one cannot fully understand how a subject comes to attribute these particular beliefs. Thus, a more complete account of mindreading must recognize the distinctive and pervasive influence of situational context in producing mental state attributions.

3.6 Assessment

To summarize the previous sections, the mindreading literature tends to focus on the cognitive processes – theorizing, simulating, etc. – that generate our attributions of beliefs to others. I have argued that the story is much more complicated, interesting, and messy than the mindreading literature lets on. Real-life social interactions are complex and dynamic, and the amount of information available swamps our cognitive processing ability. We can attend to only a tiny portion of the available information, and individuals with different social backgrounds may attend to different features of a social interaction.[16] Furthermore, people with different social backgrounds will tend to have different implicit and explicit associations and different stereotypes. They will form distinct in-groups and out-groups and thus have distinct patterns of in-group favoritism and out-group bias. Finally, they will bring different expectations to bear on social interactions. As a result, individuals with different social backgrounds often will base their mindreading judgments on different subsets of the available information. The inevitable result is that people who are situated differently in society will tend to generate different mindreading judgments, especially in ambiguous social interactions that involve people outside one's close circle of family and friends.

Of course, mindreading judgments also converge even when the mindreaders are situated differently in society. Indeed, the mindreading literature strongly focuses on convergent mindreading. The examples of convergent mindreading in the mainstream mindreading literature often are idealized and abstracted away from messy complications like those I describe in this chapter. What I offer here is a corrective to that exclusive focus on convergence. The standard mindreading story is a misleading picture of many real-life mindreading tasks, which is unfortunate because mindreading theories aim to offer descriptively accurate accounts of our ordinary social interactions. The idea that individuals will offer different interpretations of others' emotions, intentions, beliefs is completely unsurprising when one focuses on complex, realistic cases of mindreading, but it is an idea that has received very little attention in the philosophical literature on mindreading.[17]

In the next chapter, I consider how the various goals we have in a social interaction affect how we approach the task of mindreading. Most theories of mindreading presuppose that our primary goal is accurately understanding other' mental states, but I argue that we actually have lots of competing goals in a social interaction, and these different goals involve different mindreading strategies and types of explanations of behavior.

Notes

1 This chapter is built partly on previously published work. See Spaulding (2017) and Del Pinal and Spaulding (2018) for more in depth analysis of social categorization, in-group/out-group dynamics, and social biases.

2 I am not concerned in this project with reducing stereotypes, social biases, and personality traits to (sets of) beliefs. Whether or not we can reduce these cognitive phenomena to beliefs, my point is that mindreading theories do not explain how these cognitive phenomena (belief based or not) influence the process of mindreading.

3 This point is worth emphasizing as a preemptive response to the impossible and unwise advice for people to be "colorblind" (or "age-blind" or "gender-blind"). The advice is impossible because, as I go on to explain in this chapter, social categorization is automatic. Humans can't help but see others in terms of salient social groups like age, race, and gender. The advice to be "blind" to social differences also is unwise advice because it urges us to disregard others' lived experiences insofar as they differ from our own. The advice assumes that our perspectives on the world are the only legitimate perspectives. This is a bias called *Naïve Realism*, which I discuss in the next chapter. People have different experiences and perspectives in virtue of their social categories, and we learn nothing of these experiences and perspectives if we try to be blind to these differences.

4 It is an open question whether we should characterize stereotypes as beliefs or some other kind of cognitive state. I will leave this an open question and keep with the language of the empirical study of stereotypes, which regards them as belief based.

5 For comparison, babies are capable of social categorization (e.g., by forming expectations about the behavior of same-language and different-language speakers), but they are not yet capable of stereotyping. Stereotypes emerge later in the preschool years (Liberman, Woodward, and Kinzler 2017).

6 We can distinguish affective and cognitive associations. The former consists in an association between a concept with an affectively laden feature, e.g., the association between female and warm, or Black and threatening. The latter consists in an association between a concept and an affectively neutral feature, e.g., the association between old and incompetent, or Asian and being good at math. In practice, it is very likely that there is significant overlap between cognitive and affective associations (Holroyd and Sweetman 2016).

7 See Liu, Harris, and Kanwisher (2002), Forscher et al. (2016), and Bar-Anan and Nosek (2014) for more on the statistical information referenced above. Not all of these researchers are skeptical of implicit bias and its various measures, but the skeptics base their concerns on the kind of statistical information reported in these meta-analyses.

8 See Brownstein et al. (2017) for discussion of various skeptical worries about implicit bias.

9 This table is adapted from Del Pinal and Spaulding (2018).

10 This table is adapted from Del Pinal and Spaulding (2018).

11 In a review of 17 interventions for reducing implicit racial bias, Calvin Lai et al. (2014) found that the three most effective interventions for changing IAT scores involved vividly imagining counter-stereotypic scenarios, imaginatively shifting group boundaries through competition, and thinking of counter-stereotypical exemplars when taking an implicit bias test. Our model of bias predicts this kind of result. On our view, these types of interventions modulate the salient-statistical associations probed by typical measures of implicit bias, which are highly sensitive to contextual factors. Changing the context in this way *should* change the behavior of salient-statistical associations. Moreover, in a follow-up paper, Lai et al. (2016) found that even the most effective interventions on implicit bias were ineffective after delays of several hours to several days. This instability across time and contexts is exactly what our model would predict with respect to the behavior of salient-statistical associations. Thus, we take these data to confirm our conceptual model of implicit bias.

12 These remarks were made at a fundraiser on April 6, 2008.

13 These remarks were made at a fundraiser on September 9, 2016. Of course, readers also may recall that these quotes were taken out of context and both Obama and Clinton were trying to humanize conservative voters for their liberal audiences at these fundraisers.

14 The study probed the officers' *implicit* associations, but it very well could be the case that their explicit associations line up with the implicit associations. Many discussions of implicit bias contrast individuals' implicit biases with their explicit biases, but it is an open question how often implicit and explicit biases align and come apart (Holroyd 2016).

15 Think how cumbersome social interactions would be if we could not rely on the situational context, associations, and stereotypes to prime certain patterns of expectation and interpretation. We would have to consider every possible interpretation of a situation and individuals' behavior in that situation. Moreover, without associations and stereotypes, we would have little guidance determining which interpretations are more or less plausible. This would make social interactions incredibly cognitively taxing.

16 For example, if you are one of only a few people of color at a conference, race and race-related comments will be highly salient to you. However, if you are a White person at that overwhelmingly White conference, race likely will not be nearly as salient to you as some other social categories.

17 In Spaulding (2017) I consider various historical reasons why this idea has received so little attention.

References

Ames, D. R. 2004a. "Strategies for social inference: A similarity contingency model of projection and stereotyping in attribute prevalence estimates." *Journal of Personality and Social Psychology* 87 (5): 573.

Ames, D. R. 2004b. "Inside the mind reader's tool kit: Projection and stereotyping in mental state inference." *Journal of Personality and Social Psychology* 87 (3): 340–353.

Apperly, I. A., and S. A. Butterfill. 2009. "Do humans have two systems to track beliefs and belief-like states?" *Psychological Review* 116 (4): 953.

Baillargeon, R., R. Scott, and Z. He. 2010. "False-belief understanding in infants." *Trends in Cognitive Sciences* 14 (3): 110–118.

Bar-Anan, Y., and B. Nosek. 2014. "A comparative investigation of seven indirect attitude measures." *Behavior Research Methods* 46 (3): 668–688.

Bloom, P., and T. P. German. 2000. "Two reasons to abandon the false belief task as a test of theory of mind." *Cognition* 77 (1): 25–31.

Brownstein, M., N. Byrd, K. Frankish, J. Holroyd, N. Levy, E. Machery, A. Madva, S. Spaulding, and C. Sripada. 2017. "What can we learn from the Implicit Association Test? A Brains Blog roundtable." http://philosophyofbrains.com/2017/01/17/how-can-we-measure-implicit-bias-a-brains-blog-roundtable.aspx.

Carruthers, P. 2011. *The Opacity of Mind: An Integrative Theory of Self-Knowledge*. New York, NY: Oxford University Press.

Cikara, M., E. Bruneau, J. J. Van Bavel, and R. Saxe. 2014. "Their pain gives us pleasure: How intergroup dynamics shape empathic failures and counter-empathic responses." *Journal of Experimental Social Psychology* 55: 110–125.

Davies, M., and T. Stone. 1995a. *Folk Psychology: The Theory of Mind Debate*. Oxford: Blackwell.

Davies, M., and T. Stone. 1995b. *Mental Simulation: Evaluations and Applications*. Oxford: Blackwell.

DeCoster, J., and H. M. Claypool. 2004. "A meta-analysis of priming effects on impression formation supporting a general model of informational biases." *Personality and Social Psychology Review* 8 (1): 2–27.

Del Pinal, G., and S. Spaulding. 2018. "Conceptual centrality and implicit bias." *Mind & Language* 33 (1): 95–111.

Dennett, D. C. 1978. "Beliefs about beliefs [P&W, SR&B]." *Behavioral and Brain Sciences* 1 (4): 568–570.

Fiske, S. T., and S. E. Taylor. 2013. *Social Cognition: From Brains to Culture*. Thousand Oaks, CA: Sage.

Forscher, P., C. Lai, J. R. Axt, C. R. Ebersole, M. Herman, P. Devine, and B. Nosek. 2016. *A Meta-analysis of Change in Implicit Bias*. www.researchgate.net/publication/308926636_A_Meta-Analysis_of_Change_in_Implicit_Bias

Gilbert, D. T., and J. G. Hixon. 1991. "The trouble of thinking: Activation and application of stereotypic beliefs." *Journal of Personality and Social Psychology* 60 (4): 509.

Govan, C. L., and K. D. Williams. 2004. "Changing the affective valence of the stimulus items influences the IAT by re-defining the category labels." *Journal of Experimental Social Psychology* 40 (3): 357–365.

Graham, S., and B. S. Lowery. 2004. "Priming unconscious racial stereotypes about adolescent offenders." *Law and Human Behavior* 28 (5): 483.

Greenwald, A. G., D. E. McGhee, and J. L. K. Schwartz. 1998. "Measuring individual differences in implicit cognition: The implicit association test." *Journal of Personality and Social Psychology* 74 (6): 1464.

Greenwald, A. G., T. A. Poehlman, E. L. Uhlmann, and M. R. Banaji. 2009. "Understanding and using the Implicit Association Test: III. Meta-analysis of predictive validity." *Journal of Personality and Social Psychology* 97 (1): 17.

Haslam, N. 2006. "Dehumanization: An integrative review." *Personality and Social Psychology Review* 10 (3): 252–264.

Heyes, C. 2014. "False belief in infancy: A fresh look." *Developmental Science* 17 (5): 647–659. doi:10.1111/desc.12148.

Holroyd, J. 2016. "What do we want from a model of implicit cognition?" *Proceedings of the Aristotelian Society*.

Holroyd, J., and J. Sweetman. 2016. "The heterogeneity of implicit bias." In *Implicit Bias and Philosophy*, edited by Michael Brownstein and Jennifer Saul. New York, NY: Oxford University Press.

Ito, T. A., E. Thompson, and J. T. Cacioppo. 2004. "Tracking the timecourse of social perception: The effects of racial cues on event-related brain potentials." *Personality and Social Psychology Bulletin* 30 (10): 1267–1280.

Lai, C. K., A. L. Skinner, E. Cooley, S. Murrar, M. Brauer, T. Devos, J. Calanchini, Y. J. Xiao, C. Pedram, and C. K. Marshburn. 2016. "Reducing implicit racial preferences: II. Intervention effectiveness across time." *Journal of Experimental Psychology: General* 145 (8): 1001–1016.

Lai, C., M. Marini, S. A. Lehr, C. Cerruti, J. L. Shin, J. A. Joy-Gaba, A. K. Ho, B. A. Teachman, S. P. Wojcik, and S. P. Koleva. 2014. "Reducing implicit racial preferences: I. A comparative investigation of 17 interventions." *Journal of Experimental Psychology: General* 143 (4): 1765–1785.

Liberman, Z., A. L. Woodward, and K. D. Kinzler. 2017. "The origins of social categorization." *Trends in Cognitive Sciences* 21 (7): 556–568.

Linville, P. W., G. W. Fischer, and P. Salovey. 1989. "Perceived distributions of the characteristics of in-group and out-group members: Empirical evidence and a computer simulation." *Journal of Personality and Social Psychology* 57 (2): 165.

Liu, J., A. Harris, and N. Kanwisher. 2002. "Stages of processing in face perception: An MEG study." *Nature Neuroscience* 5 (9): 910–916.

Malle, B. F., and J. Holbrook. 2012. "Is there a hierarchy of social inferences? The likelihood and speed of inferring intentionality, mind, and personality." *Journal of Personality and Social Psychology* 102 (4): 661.

McClung, J. S., and S. D. Reicher. 2017. "Representing other minds: Mental state reference is moderated by group membership." *Journal of Experimental Social Psychology*.

McLoughlin, N., and H. Over. 2017. "Young children are more likely to spontaneously attribute mental states to members of their own group." *Psychological Science* 28 (10): 1503–1509.

McLoughlin, N., S. P. Tipper, and H. Over. 2017. "Young children perceive less humanness in outgroup faces." *Developmental Science*.

Mullen, B., and L. Hu. 1989. "Perceptions of ingroup and outgroup variability: A meta-analytic integration." *Basic and Applied Social Psychology* 10 (3): 233–252.

Nichols, S., and S. Stich. 2000. "A cognitive theory of pretense." *Cognition* 74 (2): 115–147.

Oberman, L. M., and V. Ramachandran. 2007. "The simulating social mind: The role of the mirror neuron system and simulation in the social and communicative deficits of autism spectrum disorders." *Psychological Bulletin* 133 (2): 310–327.

Olivola, C. Y., and A. Todorov. 2010. "Fooled by first impressions? Reexamining the diagnostic value of appearance-based inferences." *Journal of Experimental Social Psychology* 46 (2): 315–324.

Onishi, K. H., and R. Baillargeon. 2005. "Do 15-month-old infants understand false beliefs?" *Science* 308 (5719): 255–258.

Palomares, J. K. S., and A. W. Young. 2017. "Facial first impressions of partner preference traits: Trustworthiness, status, and attractiveness." *Social Psychological and Personality Science*. doi:10.1177/1948550617732388.

Premack, D., and G. Woodruff. 1978. "Does the chimpanzee have a theory of mind?" *Behavioral and Brain Sciences* 1 (4): 515–526.

Rudman, L. A., A. G. Greenwald, D. S. Mellott, and J. Schwartz. 1999. "Measuring the automatic components of prejudice: Flexibility and generality of the Implicit Association Test." *Social Cognition* 17 (4): 437–465.

Rule, N. O., N. Ambady, and R. B. Adams Jr. 2009. "Personality in perspective: Judgmental consistency across orientations of the face." *Perception* 38: 1688–1699.

Spaulding, S. 2017. "Do you see what I see? How social differences influence mindreading." *Synthese*: doi: 1-22. 10.1007/s11229-017-1404-1.

Tajfel, H. 1974. "Social identity and intergroup behaviour." *Social Science Information* 13 (2): 65–93.

Todorov, A., and J. S. Uleman. 2003. "The efficiency of binding spontaneous trait inferences to actors' faces." *Journal of Experimental Social Psychology* 39 (6): 549–562.

Trope, Y. 1986. "Identification and inferential processes in dispositional attribution." *Psychological Review* 93 (3): 239.

Vance, J. D. 2016. *Hillbilly Elegy*. New York, NY: Harper Collins.

Westra, E. 2017a. "Character and theory of mind: An integrative approach." *Philosophical Studies*.

Westra, E. 2017b. "Stereotypes, theory of mind, and the action-prediction hierarchy." *Synthese*: 1–26.

Wheeler, M. E., and S. T. Fiske. 2005. "Controlling racial prejudice: Social-cognitive goals affect amygdala and stereotype activation." *Psychological Science* 16 (1): 56–63.

Wimmer, H., and J. Perner. 1983. "Beliefs about beliefs: Representation and constraining function of wrong beliefs in young children's understanding of deception." *Cognition* 13 (1): 103–128.

Wittenbrink, B., C. M. Judd, and B. Park. 2001. "Spontaneous prejudice in context: Variability in automatically activated attitudes." *Journal of Personality and Social Psychology* 81 (5): 815.

4 The goals of mindreading

Most contemporary mindreading theories presuppose that our primary goal in mindreading is accuracy. That is, they assume that we are always aiming first and foremost to attribute the correct mental states to others. In this chapter, I argue that while this certainly is the case in some conditions, we have many different kinds of goals in social interactions and accuracy is not always a primary concern. In many other conditions our primary goals are efficiency, confirming our preexisting ideas, validating our self-worth, manipulating others, etc. I consider how the various goals we have in a social interaction affect the process of mindreading. These different goals bring about various strategies for mindreading and different types of explanations of social behavior, which are relatively unexplored by mainstream theories of mindreading.[1]

4.1 Why we mindread

A standard assumption in the mindreading literature is that the main goal we have when mindreading is to correctly attribute mental states to a target to explain and predict behavior.[2] Many works on mindreading take their *explanandum* to be *how* we accurately attribute mental states to others. This is a reasonable phenomenon to venture to explain. Considering how the theory of mind literature developed (see Chapter 3, Section 1), in particular how the false-belief task shaped subsequent discussions about theory of mind, it is no surprise that accurately attributing beliefs to others became the central focus of mindreading theories. In addition to these historical reasons, there are theoretical reasons to find accurate mental state attributions particularly worthy of study. Mental states are not observable and thus must be inferred (Spaulding 2015, 2017b), which would seem to make mindreading a difficult process. And yet mindreading seems to come quite naturally to neurotypical adult humans. Explaining this puzzling conjunction is an interesting and worthwhile endeavor.

Thus, many projects in the mindreading literature aim to explain accurate mental state attributions.

Although these are interesting, reasonable projects, the problem is that the standard assumption is false. We have various goals in social interactions, and accuracy is just one of them. Sometimes we aim for efficiency primarily and accuracy only secondarily. Other times, what matters is that we confirm our prior beliefs and values. And sometimes the goal is to shape and regulate others' mental states and behavior, rather than accurately infer their mental states (McGeer 2007; Zawidzki 2013).

Focusing exclusively on the goal of accuracy is misleading in several ways. It suggests that we often do accurately mindread, that accuracy is the only goal (worth mentioning), and that our main goal in mindreading is explaining and predicting behavior. We have good reason to doubt all three ideas. Our mindreading abilities are not nearly as impressive as suggested (Spaulding 2016). We have many different kinds of goals in our social interactions, and these goals correspond to different mindreading processes (Spaulding 2017a). And the products of mindreading differ depending on our goals and approaches to mindreading. In the rest of this chapter, I will explain these various goals, their corresponding strategies for mindreading, and the different types of explanations of social behavior that they generate.

4.2 The various goals of mindreading

Sometimes in a social interaction, what we really want to know is what the other person's mental states really are. We want to know what they are thinking, what they want, why they acted the way they did. In these cases, it matters to us that we get it right. When my best friend is acting cold and distant, I do not just want a good enough explanation of her iciness toward me. I want to know exactly why she is upset with me so that I can repair the rift. We aim for accuracy in mindreading when something important depends on getting it right, when we will be held responsible for our interpretation of the interaction, or when the situation is unusual and unexpected (Fiske and Neuberg 1990; Tetlock 1992; Kelley 1973).

Accuracy is not our only goal, though. Oftentimes we aim for efficiency primarily and accuracy only secondarily. In these cases, we want to understand what someone is thinking and why they behave in certain ways, but what we want is merely a close enough approximation of what they probably think. We do not want to devote that much time or thought to figuring it out. When the grocery store clerk seems cold and distant, I am mildly curious about the icy attitude but not so curious that I carefully deliberate about it. After all, store clerks often are impersonal and cold. When the social interaction seems ordinary and familiar, when not much hangs on it, or when we are otherwise cognitively taxed, we use cognitive shortcuts, e.g., stereotypes and projection (Fiske and Taylor 2013, 177–199).

Sometimes our goals in a social interaction have more to do with *us* than understanding others' mental states and behavior. We have all witnessed people clearly seeing what they want to see in a social interaction, e.g., when a jealous lover cannot help but see deception in their partner's behavior, or when a racist person always infers sinister motives in people of another race, or (for a more positive example) when a grandparent sees only laudable intentions of a beloved grandchild.[3] In these cases, anxiety reduction, self-esteem protection, and confirmation of one's opinions guide our inferences (Dunning 1999; Kunda 1990).

Finally, in some cases our goal in attributing mental states is to regulate others' minds and behavior. In these cases, you aim to make it the case that you or others have the mental states you attribute, even if they do not decisively have those mental states when you attribute them. For example, when I tell myself or say aloud to my husband that I want to go for a run in the morning, I am in no small part trying to get myself to adopt an intention I regard as good and to be the kind of person who acts on her good intentions. When you tell an underachieving student that you know he wants to do well in the class, part of what you are doing is encouraging him to adopt that desire and act accordingly. In these sorts of cases, our primary goal in mindreading is not to explain and predict behavior. The aim is to shape and regulate behavior (McGeer 2007; Zawidzki 2013).

The goals described above are not mutually exclusive and probably not exhaustive of all of our goals in social interactions. We may have several goals in a given social interaction, though to varying degrees and with respect to different aspects of the interaction. So, for example, in mindreading my spouse, I may want to figure out what he truly thinks about, say, the recent wave of sexual harassment allegations against some politicians, but at the same time I may want to shape what he is thinking to be more in line with what I think about these allegations. My goal to figure out what he thinks pulls in a different direction from my goals to shape what he thinks and to confirm my own perspective. How I approach this interaction will depend on which of these is my primary goal. Moreover, my goals can change within the interaction and some goals that were subsidiary can become primary. At first I may be mostly curious about what my husband thinks about the sexual harassment allegations. But if it starts to seem to me that he has a *different* view on these allegations than I do, the priority of my goals may start to shift as the conversation proceeds and I may engage in less accuracy-oriented mindreading and more mind *shaping*. This fictional but realistic example illustrates that social interactions are complex, dynamic affairs.

4.3 Approaches to mindreading

Corresponding to this array of goals are various mindreading processes.[4] When our primary goal is to correctly understand someone's mental states, we tend to search for information relevant to that person's mental states in a controlled and deliberative fashion. For instance, when I am interviewing potential nannies, it is really important to me that I accurately mindread the interviewees. I want to know whether a potential nanny likes my daughter or is just being friendly to get a job, how she will respond to various requests and potentially criticism, how committed she is to taking and keeping this job, whether she is patient and reliable, etc. I want to ensure that my judgments about the potential nannies are accurate because I am entrusting this person to care for my daughter. This kind of reasoning is effortful, cognitively taxing, and difficult if one is under cognitive load or not well practiced in this kind of reflective reasoning (Gilbert, Krull, and Pelham 1988). People may make mistakes in accuracy-oriented mindreading when the information they gather is too limited (i.e., small sample size), their information is distorted, or there are extreme outliers that inappropriately skew the sample. Even when people have accurate information, they may make mistakes when they are under increased cognitive load. When they are stressed, tired, or distracted they may pay attention to superficial but misleading cues and not attend to relevant but inconspicuous information. They may also make mistakes when their information search is skewed by motivated reasoning, e.g., when they are subject to confirmation bias.

We may take a deliberative approach to mindreading when we have mind shaping or mind regulative goals, as well. When we want to encourage someone to have and act on a certain belief, desire, or goal, we may carefully examine and present evidence to the target. Most mindreading is interactive in the sense that mindreaders engage with the target and environment in the process of mindreading,[5] but the approach we take when we are mind shaping or regulating is more purposefully interactive. We carefully choose to curate the interaction so that the target will adopt a mental state we prefer them to have. We may not always be fully aware that we are mind shaping because introspection is notoriously bad at revealing why we are behaving in a certain way (Nisbett and Wilson 1977). Nevertheless, we may adopt this deliberative approach even when we do not clearly understand our own motives.

When our motivation for efficiency is stronger than our motivation for accuracy, several mindreading approaches are available. Which approach we take depends on how we socially categorize individuals we are mindreading.[6] When we perceive an individual to be *similar* to us in some salient respect, that is, when we regard that person as part of a relevant in-group, the efficient approach to mindreading is to simply project our own mental states to that individual (Ames 2004a, b). Relatedly, we sometimes use our mental states as an anchor and adjust the interpretation based on how similar the individual is to us. These egocentric heuristics are efficient shortcuts for getting a good-enough approximation of what others think. Importantly, given the primary goal of efficiency, the process of mindreading is different. Again, the claim is not just that what we do with our mindreading judgments differs depending on our goals. Rather, how we go about mindreading – projecting or stereotyping vs. carefully culling all the relevant information – is dramatically different depending on our goals.

The success of these efficient mindreading approaches depends on whether we have correctly judged the similarity between the target and us and whether we have correctly introspected our own mental states. Thus, there are at least three ways in which this approach can lead to error: we could overestimate the similarity between the target and us and inappropriately project our own mental states, we could overestimate the similarity and insufficiently adjust the projection to the target's situation, and we could correctly judge the similarity but be mistaken about what our own mental states actually are.[7]

When we perceive an individual to be *different* from us, our efficient approach to mindreading that individual involves stereotypes about the individual's salient in-group (Krueger 1998; Vorauer et al. 2000; Ames 2004b). As I explain in Chapter 3, I take stereotypes to be positive, negative, or neutral beliefs about members of a social group. Some stereotypes may involve beliefs about the mental states of members of certain social groups, and we may directly attribute a stereotype about the mental states of certain social groups. Stereotypes may involve beliefs about a target group's mental states or beliefs about a target group that imply certain kinds of mental states in certain contexts. For example, a common stereotype is that women care more about family than careers. Suppose that a faculty member notices that a female colleague who also happens to be a mother of a young child has produced less research since the child was born and muses to himself that her decreased productivity probably is due to her wanting to spend more time with her family than on her career. She is a mom now after all, he observes. When he explains his colleague's pattern of behavior, he is simply attributing to her the stereotype that women care less about their careers than about their families. Stereotypes need not involve mental states at all, of course, but the kind of stereotypes I am interested in here do. The example above is a case where the stereotype involves an assumption about what a target's mental states are.[8]

Stereotype-based mindreading is an efficient cognitive approach because all one needs to do is retrieve a stereotype from long-term memory rather than deliberate about what may have caused someone to act as they have.

We can simply attribute a whole cluster of related beliefs, desires, values, personality traits, etc. to a target without having to contemplate what a target actually believes, desires, etc.[9] In this way, stereotypes reduce cognitive load. Moreover, once the stereotype is activated, processing stereotype-consistent information is less cognitively demanding than processing stereotype-inconsistent information. When subjects are under cognitive load, they will use the most efficient approaches, e.g., employing a stereotype and attending to stereotype-inconsistent information only if it is highly salient (Gilbert and Hixon 1991).

Stereotype-based mindreading can go awry in several ways. First, we could underestimate the similarity between the target and us and baselessly apply a stereotype when projection would be a more appropriate heuristic. In these cases, we fail to appreciate that we are more similar to the target in the relevant respects than we assumed. Alternatively, we could correctly diagnose the difference between us and the target but make mindreading mistakes because we employ a false or inappropriate stereotype. Some stereotypes involve false beliefs about members of social groups, which could distort the inference about the mental states of a member of that social group. For example, a common but false stereotypical belief is that disabled people have a lower quality of life than non-disabled people. In fact, able-bodied and disabled people report approximately the same levels satisfaction with their lives (Albrecht and Devlieger 1999). Disabled people are not more likely to have lower quality of life, nor is lower quality of life a characteristic or distinctive property of disabled people. That is, it is not the case that *most* disabled people have lower quality of life than able-bodied people, nor is it the case that *more* disabled people than able-bodied people have lower quality of life, and knowing just that someone has a lower quality of life does not give us any indication about whether that person is disabled. With respect to mindreading, if one attributed this false stereotype to a disabled individual, then one would be more likely to infer depressed, negative, cynical mental states.

We should be careful not to conflate false or inappropriate stereotypes, like the one described above, from stereotypes that express generics. Generics often are technically statistically incorrect, but they are not false or inappropriate in the way that some stereotypes are. Generics involve the attribution of a common characteristic or striking property to a kind whose members are disposed to manifest that property (Leslie 2008). *Many* stereotypes express generics. Consider the stereotype that mosquitoes carry West Nile virus. Statistically speaking, this is false. Not all mosquitoes carry West Nile virus. Indeed, the vast majority of mosquitoes do *not* carry the virus. Nevertheless, we take the statement "mosquitoes carry West Nile virus" to be true. Why? If all you knew about an animal was that it carried

West Nile virus, you would have very good reason to think that animal was a mosquito. So, the belief that mosquitos carry West Nile virus expresses what we take to be a striking property of mosquitoes. It is a generic, not a false stereotype.

Let's consider a stereotype that is more relevant to mindreading: the stereotype that women are family oriented, as opposed to career oriented. That is, women *desire*, *value*, or *care* more about their families than their careers. This stereotype expresses a generic. Not *every* woman is family oriented, and people who hold the stereotype that women are family oriented probably would not believe that *every* woman is family oriented. So, what information is a stereotypical generic encoding? The informational structure of the stereotypical generic could take different forms.

(i) The majority of women prioritize family over career.
(ii) When we think of women, we think of family.
(iii) Compared to men, women tend to be proportionally more family oriented (even if most of them are not family oriented).
(iv) Prioritizing family explains other typical properties of women.

The logical form of each of these alternatives is different: (i) is about the distribution of family orientation amongst women, (ii) is about the prominence or salience of family orientation in our conception of women, (iii) is about the cue validity of family orientation for identifying gender, and (iv) is about the explanatory role and hence centrality of family orientation in explaining other typical properties of women (Del Pinal and Spaulding 2018).[10]

The point here is that false or inappropriate stereotypes are different from stereotypes that express generics. Most stereotypes express generics, and these generics may have different underlying structures. These generic-expressing stereotypes may still lead to mindreading errors, but not quite in the same way that false or inappropriate stereotypes do.

I have examined several forms of efficient mindreading and the ways in which they may lead to error. Whereas we tend to deliberate carefully about a target's mental states when a lot hangs on our getting it right, we will be held responsible for the effects of our judgments, we care personally about getting it right, or when the situation is unusual or unexpected, we tend to use efficient mindreading approaches like stereotyping and projection when these conditions do not hold, especially when we are in familiar situations. Indeed, as particular situations become more familiar to us, and we use these shortcuts more regularly, the projective or stereotypical interpretation of familiar situations becomes more accessible and more difficult to override (Higgins, King, and Mavin 1982). The tendency to habitually code situations and others' behavior in a particular way can become proceduralized, which makes social interpretation easier, more efficient, and more predictable, but it also preempts equally reasonable but less practiced judgments (Smith 1990). Thus, in very familiar situations it may be difficult for people *not* to employ particular stereotypes or project their own perspective.

I have discussed the mindreading approaches when our primary goals are accuracy, mind shaping, and efficiency. The approaches to mindreading we take when we have self-serving goals may be effortful or efficient, and we likely have little insight when we are adopting these self-serving approaches to mindreading. In the remainder of this section, I shall discuss several common patterns of motivated reasoning that serve self-interested goals described above.

The Self-Serving Attributional Bias describes our tendency to take credit for success and deny responsibility for failure (Miller and Ross 1975).[11] We often attribute our successes to some internal factor, e.g., our diligence or talent, and attribute our failures to external mitigating factors, e.g., bad luck or bias against us. In this way, we come to feel good about our successes and brush off our failures. This pattern of reasoning generally is a sign of healthy psychological function.

This pattern of reasoning is found for judging in-group and out-group behaviors, as well. This is called the Group-Serving Attributional Bias (Brewer and Brown 1998; Pettigrew 1979). One tends to judge the *success* of an *out-group* to be the result of situational factors outside of their control, such as good luck and bias for them, and one judges the *failure* of an *out-group* to be the result of internal flaws of character, intelligence, or talent. In contrast, one judges the *success* of one's *in-group* to be the result of internal strengths of character, intelligence, and talent, and one judges the *failure* of one's *in-group* to be the result of situational factors outside of their control, such as bad luck and bias against them.

This is a clunky description of the bias, but put in terms of sports it is quite intuitive. Even though I know about this bias, I cannot help but fall prey to this reasoning when watching the Badgers play. When the Badgers win, I chalk this up to their talent and hard work. But on the rare occasion that the Badgers lose, it clearly is because they had a bad night, they had a few unlucky injuries, the other team got some lucky breaks, and the referees were biased against my team on some crucial plays. This sounds silly in the abstract, but it is so utterly common and difficult to resist.

We explain a target's behavior differently depending on whether the target is part of our in-group or an out-group. This differential pattern of explanation serves to protect one's self-esteem in both the individual and group case. Recall that we tend to like and prefer our in-groups more than out-groups. The group-serving attributional bias provides (circular) justification for these feelings and for our charitable beliefs about our in-group and our uncharitable beliefs about out-groups. The Self- and Group-Serving Attributional Biases are thus one type of mindreading approach we take when we have self-serving goals.

Naïve Realism is another bias that we are subject to when self-serving goals motivate our mindreading. Naïve Realism is the tendency to regard others as more susceptible to bias and misperception than oneself (Pronin, Lin, and Ross 2002; Pronin 2007). We think we simply see things as they are, but others suffer from bias. Our explanations of others' views focus on personal motivation, misperceptions, and skewed reasoning, whereas we regard ourselves and those who agree with us as simply correct. This tendency is especially prevalent in contexts of competition or disagreement, e.g., sports and politics (Kennedy and Pronin 2008). This bias influences how we interpret others' mental states and behavior, especially out-group members who disagree with us about some salient issue, and it serves to confirm our own perspectives and justify our dismissal of views that contradict our own. This is another example of how we engage in motivated reasoning about others' mental states and behavior when self-serving goals influence us.

Both Naïve Realism and the Self- and Group-Serving Attributional Biases further our self-serving goals through confirmation bias, which is our tendency to notice and consider only information that confirms our preconceived ideas, prior knowledge, and relevant expertise and to interpret ambiguous information in light of this prior information. As we saw in Chapter 3, social categorization, along with implicit and explicit learning, and our tendency for in-group favoritism yield more charitable, friendly interpretations of the mental states and behavior of those we regard as similar to us. These processes also generate less charitable, sometimes neutral or even malicious interpretations of the mental states and behavior of those we perceive to be unlike us. In other words, these biases serve to circularly justify beliefs and preferences that come about from our perceived similarity judgments and a tendency to like people who are similar to us. These biases affect both deliberative, controlled processes and efficient processes like projection and stereotyping. Confirmation bias may occur even when the source of the information is known to be unreliable, the claims are highly improbable, and even when accuracy is incentivized (Skov and Sherman 1986; Slowiaczek et al. 1992; Snyder, Campbell, and Preston 1982).

This is not to say that we *always* get confirmation of our world view. I am drawing a cynical picture of our social interactions here, but it is not *that* cynical. Some factual information may clearly contradict our preconceived ideas, and we have to reconcile the information with our ideas (Wood and

Porter 2018). Encountering factual information that contradicts our beliefs may lead us to make different mindreading inferences, revise our stereotypes, or at least make exceptions to our stereotypes. However, social interactions can be ambiguous, we may not have access to hard and fast facts, and the interpretation of these ambiguous interactions differs greatly for individuals who have very different experiences and preconceived ideas. Thus, though facts matter, when we are called to interpret an ambiguous social interaction, we tend to prioritize information that confirms our ideas.

To complete our examination of the prospects for success of the various mindreading approaches, it is not difficult to see how the biases that we employ when we have self-serving motivations might lead to mistaken mindreading judgments. In fact, if we get it right when we are subject to the Self/Group-Serving Attributional Biases and Naïve Realism, it seems that this would be a matter of luck as these biases distort our information gathering and reasoning processes. Of course, things become much more complicated when we display our biased reasoning to the targets, because in some cases the targets may conform to the mental states and features attributed to them. This is allegedly what happens with stereotype threat and projections of gender stereotypes on infants and children (Haslanger 2007; Mameli 2001; Spencer, Logel, and Davies 2016). In such cases, what is initially an incorrect inference eventually becomes correct. These are examples of unintentional mind shaping.

To recap the last two sections, Sections 3.2 and 3.3 explain how our goals in a social interaction influence the type of cognitive approach to mindreading we adopt and the ways in which these processes go awry. When we want to accurately understand others' mental states, we aim to exhaustively review the available social information and attribute mental states to others in that way. This is also one way in which we engage in purposefully mind shaping other people. When we lack the motivation or ability to do an exhaustive search, we take shortcuts that are a function of social categorization. We are more likely to project our own mental states on those we perceive to be similar to us and stereotype those we perceive to be different from us. In addition, many of our social interpretations are guided by self-serving goals, and as a result we interpret others' behavior and mental states in light of what we antecedently believe or want to believe.

This is a good place to reiterate that the goals in our social interactions are not mutually exclusive. We may have multiple goals, and these goals may shift in priority as the interaction proceeds. As a result, we may adopt multiple approaches to mindreading, which may also shift and interact as the social interaction proceeds. Deliberation can influence and be influenced by heuristics and self-serving biases. Moreover, we may lack insight into the goals and approaches we adopt to serve those goals. Many of our social interactions proceed instinctively. We often do not reflect on why we behave as we do, and even when we do so reflect, we may lack accurate insight into our motives.

4.4 Mindreading products

Whereas the previous two sections focus on our goals and approaches to mindreading, this section focuses on the products of mindreading. It is typically assumed that there is just one sort of mindreading product: a judgment about a target's mental states that explains and predicts the target's behavior. I argue in this section that this standard assumption glosses over interesting differences in the kinds of explanations we generate in mindreading. It turns out that the products of mindreading are as varied as the goals and approaches we take to mindreading.

Mindreading generates explanations that involve mental states. In this section, I will explore the nature of these explanations. Though there are many different ways to categorize explanations, I will use the following taxonomy, derived from Aristotle's four causes and fruitfully employed in the psychological literature: teleological explanations, mechanistic explanations, and formal explanations (Lombrozo 2012; Lombrozo and Carey 2006; Dennett 1987). Teleological explanations cite functions or goals. For example, one might explain a cat chasing a mouse by saying that the cat wanted to eat the mouse. Mechanistic explanations cite proximal causes. In explaining the cat chasing the mouse, one might say that chasing the mouse stimulates the release of dopamine in the cat's brain. Formal explanations cite kind membership and norms. A formal explanation of the cat's behavior would say that this is just what cats do; they chase mice.

Teleological and mechanistic explanations clearly are causal explanations. Explaining the cat's behavior by citing the hormones released in its brain is an explanation at the neurological level of what causes the cat to chase the mouse. Citing the cat's desires or intentions to chase the mouse explains what causes the behavior at a psychological or folk psychological level. The nature of formal explanation is up for debate, however the consensus is that formal explanation is non-causal (Lombrozo 2012; Prasada and Dillingham 2006, 2009). On this view, you explain something by categorizing it as part of a larger class of phenomena. For example, you explain why an animal has four legs by citing the fact that it is a dog and dogs are the kind of creatures that typically have four legs.

The issue of whether formal explanations are causal will turn out to be relevant to this project, so I will spend a bit of time on the issue here. Although the non-causal interpretation is the consensus view, I think there is good reason to hold that formal explanations at least *can* be causal. Formal explanations often seem compelling only when we import causal assumptions about the relationship posited in the explanation. When I tell you that an animal has four legs because it is a dog, unless you assume that being a dog (i.e., having the genetic code of a dog) is partly causally responsible for this animal having four legs, the explanation will seem useless to you. Otherwise, I am simply telling you something you already know – that this is a dog and dogs have four legs. In some cases, we seem to want more than just classificatory information out of formal explanations. Indeed, Wilkenfeld, Gleason, and Lombrozo (Forthcoming) find that people will infer causal information about categories even when you explicitly instruct them not to do so. Moreover, they find that whether we find diagnostic categories explanatory depends, at least in part, on whether there is a causal basis for category membership. This is further evidence that formal explanations are seen as explanatory only when we import causal assumptions.

Consider a stereotypical explanation of behavior, which seems to be an example of formal explanation. When the faculty member explains that his colleague is less productive because she is a mom now, he is explaining her behavior by categorizing it as an instance of a familiar pattern of behavior, namely, women putting their careers on the back burner to focus on their families. This does not seem to be a mechanistic or teleological explanation. The explanation does not cite proximal causes of behavior, and he is not implying that the function of mothers is to value their family over their careers. (This would be an unusual and inappropriate sense of *function*.) Rather, the explanation seems to be a clear-cut example of formal explanation, i.e., of categorizing an instance of behavior as a part of a larger pattern of behavior.[12] The question is whether this explanation is causal. If this explanation were merely categorical, it would seem to be provide no useful or interesting information because the fact of her motherhood is not new information. If one did not assume that the mom category has particular causal basis that causes characteristic properties, then it would be an uninformative explanation. The statement, "She's a mom now, after all" could be elliptical for many things, e.g., that moms care less about their careers when they have a family, or that moms tend to be the primary caretakers of young children and therefore have no choice but to put their children first, that (in the United States, at least) maternity leave and daycare options are terrible and thus necessitate mothers sacrificing their careers if they want to have families. The faculty member could have assumed any of these other background ideas in asserting that the less productive colleague is a mom now. The point is that these are all causal assumptions about the relationship between the category and the feature, and without some kind of assumption like this, the explanation would ring hollow.

The view that formal explanation is causal is an unconventional view, but not outlandish. After all, the argument that teleological or functional explanations are causal runs in roughly the same way. People judge teleological and functional explanations as acceptable explanations only when they reference functions that play a predictable causal role (Lombrozo and Carey 2006). Teleological or functional explanations that do not meet this condition are regarded as not explanatory. We tend to regard those explanations as circular or uninformative. For reasons such as these, many theorists have concluded that teleological explanations often presuppose causal information (Lombrozo 2012; Piaget 1969; Gelman 2003, 2004). The same reasoning applies to formal explanations.

With the digression about the nature of formal explanation complete, we can now return to the kinds of explanation we generate in different contexts. In general, the kind of explanation you generate or find satisfying will depend on your goals and the context. Tania Lombrozo and Susan Carey convincingly argue that the psychological function of explanation is to provide information that can be exported to novel cases (Lombrozo and Carey 2006).[13] Their hypothesis, called *Explanation for Export*, holds that we prefer explanations that provide information that we can use in generalizations, predictions, and future interventions. The Explanation for Export hypothesis neatly explains data on what kinds of explanations people accept, e.g., teleological explanations only in which the function cited plays a causal role via a causal process that conforms to a predictable pattern (Lombrozo and Carey 2006, 196).[14]

People prefer explanations that highlight information that will be useful for inductive inferences they expect to make (Vasilyeva, Wilkenfeld, and Lombrozo 2017). For example, when the task at hand involves identifying the function or goal of some thing or behavior, teleological explanations are regarded as better than other kinds of explanations. Teleological explanations seem to be the default as they come most naturally to children and adults (Lombrozo 2012). Presumably this is because we find functions and goals intuitively useful for making generalizations, predictions, and interventions.[15] In contrast, when the task involves inferring the causal origin of some thing or behavior, mechanistic explanations are regarded as more appropriate. In these cases, information about proximal causes provides useful information for future inferences and interventions. In other cases, when we are interested in categorizing things or behaviors, formal explanations are most appropriate.

With respect to mindreading, teleological explanations are the default (as they are in general). When we want to predict or expect to intervene on someone's behavior, understanding their goals is extremely useful.[16] This should come as no surprise to anyone with even passing familiarity with the mindreading literature. What is new here is that we will sometimes generate different kinds of explanations, depending on our goals and the context.

We sometimes offer stereotypical explanations rather than teleological explanations. Stereotypical explanations, I argue above, are a kind of formal explanation. Stereotypical formal explanations interpret a target's behavior in terms of a more general category, which allows us to make predictions about future behavior. We generate or accept these explanations when we are motivated more by efficiency than accuracy, when the person we are mindreading is part of the out-group, and when we are more interested in categorizing the behavior as part of a more general pattern than intervening on an individual's future behavior. We may be interested in categorization when we are moralizing about a behavior, social signaling, or protecting our self/in-group image, motivations that are not mutually exclusive. These stereotypical explanations are compelling only when they presume causal information connecting the behavior and social category stereotype. An example will help illustrate this kind of case. When the faculty member notices that his colleague publishes less now that she is a mother, and when he is more interested in a pithy explanation than a careful analysis of her motives and situation, he might offer the stereotypical explanation that she is a mom now and so she puts her career on the back burner. This explanation may serve to protect his ego as he regards his colleague as a competitor in terms of research productivity, and this is evidence that he is in fact a more dedicated and hardworking researcher. The stereotypical explanation can only protect his ego if we assume that there is a causal connection between becoming a mother and being less productive, e.g., that mothers tend to value their family more than their career. Otherwise, the explanation does not confirm to the faculty member that he is more dedicated and hardworking; it is just an uninformative remark.[17]

The *explanandum* of mindreading may be a target's behavior – as many of my examples show – but the *explanandum* also could be the target's mental states. In such a case, one might be interested in explaining why a target has particular mental states. Teleological and formal mindreading explanations could explain either behavior or mental states. For teleological explanations, we may want to figure out the reasons why a target behaves a certain way, but we may also want to know why a target has certain beliefs, desires, emotions, intentions, etc. For formal explanations, we may want to categorize behavioral patterns as typical of certain social groups, but we may also want to categorize a target's mental states as typical of certain social groups. In contrast to teleological and formal explanation, mechanistic mindreading explanations are best suited to explain mental states. Mechanistic mindreading explanations aim to explain mental states by reference to proximal causes of those mental states, which may include particular neurological causes, other mental states, sensations, the perceptual environment, etc. Though there is no relevant empirical data on the prevalence of mechanistic mindreading explanations, it is reasonable to speculate that these are less common than either teleological or formal mindreading explanations.

4.5 Assessment

This is a long chapter with a lot of information, so it is worth giving a brief summary of the terrain I have covered in this chapter. The mainstream mindreading literature implicitly assumes that our goal in mindreading is to accurately attribute mental states to a target and that the function of such attributions is to generate teleological explanations of behavior. Though we sometimes do have that kind of goal, this view captures only a very small piece of mindreading. There is much more variety and nuance in our day-to-day mindreading practices.

We may have an assortment of goals in mindreading, including accuracy, social manipulation, efficiency, and self-interest. We may adopt more than one of these goals, and the priority of our various goals may shift as the social interaction proceeds. These goals correspond to distinctive strategies for mindreading, including careful deliberation, heuristics like projecting and stereotyping (depending on whether or not the target is part of a relevant in-group), and various kinds of motivated reasoning. We may adopt multiple mindreading strategies, which may also shift and influence the implementation of other strategies over the course of the interaction. Finally, the mindreading explanations we produce vary by our goals and the context of mindreading. We often produce teleological mindreading explanations, which serve certain kinds of inductive functions. But we may also produce formal and mechanistic mindreading explanations, which serve different kinds of inductive functions.

Thus, to summarize, our goals, the mindreading approaches we take that correspond to these goals, and the explanations we produce as a result of our goals and approaches all can vary. The idea that there is just one goal, one approach, and one kind of product of mindreading worth discussing underestimates the diversity, messiness, and interestingness of our mindreading practices.

Notes

1 This chapter builds on parts of my previous work. See Spaulding (2017a, 2016).

2 It is an open question whether this is the *function* of mindreading. The function of mindreading *could* be accurately explaining and predicting behavior. Or it could be enabling cooperation, allowing us to manipulate the social world, or something else altogether. I take it that questions about the function of mindreading will involve evolutionary speculation about how mindreading developed and the problems it

solved. I try not to engage in evolutionary speculations (though see Chapter 7 for some minimal evolutionary speculation), so I leave it to others to figure out what the function of mindreading is. See Heyes and Frith (2014), Mameli (2001), and Carruthers (2009) for different evolutionary hypotheses about mindreading.

3 We often see these inferential patterns in others, but we are not very good at identifying our own self-motivated reasoning. This is a bias called *Naïve Realism*, which I will discuss in the next section.

4 These goals and corresponding mindreading approaches may be conscious and deliberate, but for the most part they need not be. Like the goals that they correspond to, these cognitive approaches are not mutually exclusive. A given social interaction may involve bits of each cognitive approach, relative to the priority of one's goals, and the various cognitive approaches may interact and sometimes interfere with each other.

5 Although enactive and embodied cognition theorists are right to object that this interactive element has been underexplored in the mainstream mindreading literature.

6 See Chapter 3 for more on social categorization.

7 The effects of the first two kinds of mistake are called *the curse of knowledge*, a phenomenon wherein we falsely assume that others know what we know, and *the false consensus effect*, which occurs when we falsely assume that others share our opinion on some matter (Clement and Krueger 2002; Epley and Waytz 2010, 512).

8 Stereotypes also play a subtler role in mindreading. The stereotype may be activated (based on the kind of social categorization described in Chapter 3), and the salience of the stereotype influences our attribution of traits, mental states, and how we interpret behavior. This is not necessarily an efficient cognitive approach to mindreading. Rather, this a way in which stereotypes could influence any approach to mindreading.

9 See Westra (2017) for an interesting and plausible account of how we employ stereotypes in our theory of mind predictions.

10 If we want to figure out the nature of a subject's stereotypical belief – and perhaps intervene to present counter information – we will use different sorts of tests. For example, a subject with belief (i) is more likely to be surprised with evidence that large numbers of women are career oriented than someone with belief (ii). Similarly, presenting someone with relative rates of, e.g., men and women taking a more flexible job to help care for family, will bear directly on (iii) but have little bearing on (i) or (ii). Finally, asking subjects how central family orientation is to their conception of women – by measuring how surprised they would be to learn of a woman who was not family oriented vs. how surprised they would be to learn of a man who was not family oriented, or by having them list other typical characteristics of men and women and map out which features explain other features – bears on (iv), but not the other kinds of beliefs.

11 This bias differs from the more well-known Actor-Observer Effect, which holds that people explain others' behavior in terms of dispositional factors and their own in terms of situational factors. In other words, behavioral explanations differ depending on whether one is the actor or the observer. The empirical evidence for this effect is mixed (Malle 2006), but see Malle, Knobe, and Nelson (2007) for a novel interpretation of the asymmetries in patterns of explanation.

12 We could also interpret this explanation as categorizing the behavior in terms of motherhood norms. Social role concepts often involve norms for behavior (Kalish and Lawson 2008). It seems that explaining a phenomenon by referring to a norm is a kind of formal explanation, but it may very well be the case that

normative explanations require their own category. I leave this question open for future work on taxonomies of social explanations.

13 Lombrozo (2012) clarifies that explanations may not all have the same function (see, for instance, rationalizing explanations), and it is unlikely that individuals have the explicit goal of fulfilling the function of the type of explanation they are generating or accepting.

14 The Explanation for Export hypothesis also makes sense of the idea that formal explanations are, or can be, causal. Typically, causal information is crucial for making generalizations and interventions. If this is right, then we would find formal explanations compelling and useful only when we import causal assumptions that are potentially useful for generalizations, predictions, and future interventions. See Wilkenfeld, Gleason, and Lombrozo (Forthcoming) for further empirical evidence of this idea.

15 It is put in different terms, but this is essentially Dennett's (1987) view about the intentional stance.

16 Most theorists working in this area presuppose that mindreading explanations are causal, e.g., the cause of Jane's going to the cupboard is her belief that there are cookies in the cupboard and her desire to eat cookies. Bertram Malle (2004) argues that explanations that cite an agent's reasons are importantly different from causal explanations. If he is right, then it seems that we need another category for reasons explanations. In my view, mindreading explanations do have causal elements to them, so it is important to regard them as causal even if they may also serve to rationalize behavior. Malle is particularly concerned to distinguish *mechanistic* causal explanations from reasons explanations. He is right that these are importantly different kinds of explanations, but we can recognize that while also holding that reasons explanations are causal.

17 Given the prevalence of this extended example, it is worth remarking that this is not a true story. It is merely inspired by actual events, as they say in films.

References

Albrecht, G. L., and P. J. Devlieger. 1999. "The disability paradox: High quality of life against all odds." *Social Science & Medicine* 48 (8): 977–988.

Ames, D. R. 2004a. "Strategies for social inference: A similarity contingency model of projection and stereotyping in attribute prevalence estimates." *Journal of Personality and Social Psychology* 87 (5): 573.

Ames, D. R. 2004b. "Inside the mind reader's tool kit: Projection and stereotyping in mental state inference." *Journal of Personality and Social Psychology* 87 (3): 340–353.

Brewer, M. B., and R. J. Brown. 1998. "Intergroup relations." In *The Handbook of Social Psychology*, edited by D. T. Gilbert, S. T. Fiske and G. Lindzey, 554–594. New York, NY: McGraw-Hill.

Carruthers, P. 2009. "How we know our own minds: The relationship between mindreading and metacognition." *Behavioral and Brain Sciences* 32 (2): 1–18.

Clement, R. W., and J. Krueger. 2002. "Social categorization moderates social projection." *Journal of Experimental Social Psychology* 38 (3): 219–231.

Dennett, D. C. 1987. *The Intentional Stance*. Cambridge, MA: MIT Press.

Dunning, D. 1999. "A newer look: Motivated social cognition and the schematic representation of social concepts." *Psychological Inquiry* 10 (1): 1–11.

Epley, N., and A. Waytz. 2010. "Mind perception." In *Handbook of Social Psychology*, edited by Susan T. Fiske, Daniel T. Gilbert and G. Lindzey, 498–451. Hoboken, NJ: Wiley.

Fiske, S. T., and S. L. Neuberg. 1990. "A continuum of impression formation, from category-based to individuating processes: Influences of information and motivation on attention and interpretation." *Advances in Experimental Social Psychology* 23: 1–74.

Fiske, S. T., and S. E. Taylor. 2013. *Social Cognition: From Brains to Culture*. Thousand Oaks, CA: Sage.

Gelman, S. A. 2003. *The Essential Child: Origins of Essentialism in Everyday Thought*. Oxford: Oxford Series in Cognitive Development.

Gelman, S. A. 2004. "Psychological essentialism in children." *Trends in Cognitive Sciences* 8 (9): 404–409.

Gilbert, D. T., and J. G. Hixon. 1991. "The trouble of thinking: Activation and application of stereotypic beliefs." *Journal of Personality and Social Psychology* 60 (4): 509.

Gilbert, D. T., D. S. Krull, and B. W. Pelham. 1988. "Of thoughts unspoken: Social inference and the self-regulation of behavior." *Journal of Personality and Social Psychology* 55 (5): 685.Haslanger, S. 2007. "'But mom, crop-tops are cute!' Social knowledge, social structure and ideology critique." *Philosophical Issues* 17 (1): 70–91.

Heyes, C. M., and C. D. Frith. 2014. "The cultural evolution of mind reading." *Science* 344 (6190): 1243091.

Higgins, E. T., G. A. King, and G. H. Mavin. 1982. "Individual construct accessibility and subjective impressions and recall." *Journal of Personality and Social Psychology* 43 (1): 35.

Kalish, C. W., and C. A. Lawson. 2008. "Development of social category representations: Early appreciation of roles and deontic relations." *Child Development* 79 (3): 577–593.

Kelley, H. H. 1973. "The processes of causal attribution." *American Psychologist* 28 (2): 107.

Kennedy, K. A., and E. Pronin. 2008. "When disagreement gets ugly: Perceptions of bias and the escalation of conflict." *Personality and Social Psychology Bulletin* 34 (6): 833–848.

Krueger, J. 1998. "On the perception of social consensus." *Advances in Experimental Social Psychology* 30: 164–240.

Kunda, Z. 1990. "The case for motivated reasoning." *Psychological Bulletin* 108 (3): 480.

Leslie, Sarah-Jane. 2008. "Generics: Cognition and acquisition." *Philosophical Review* 117 (1): 1–47.

Lombrozo, T. 2012. "Explanation and abductive inference." In *Oxford Handbook of Thinking and Reasoning*, edited by Keith J. Holyoak and Robert G. Morrison, 260–276. Oxford: Oxford University Press.

Lombrozo, T., and S. Carey. 2006. "Functional explanation and the function of explanation." *Cognition* 99 (2): 167–204.

Malle, B. F. 2004. *How the Mind Explains Behavior: Folk Explanations, Meaning, and Social Interaction*. Cambridge, MA: MIT Press.

Malle, B. F. 2006. "The actor-observer asymmetry in attribution: A (surprising) meta-analysis." *Psychological Bulletin* 132 (6): 895.

Malle, B. F., J. M. Knobe, and S. E. Nelson. 2007. "Actor-observer asymmetries in explanations of behavior: New answers to an old question." *Journal of Personality and Social Psychology* 93 (4): 491.

Mameli, M. 2001. "Mindreading, mindshaping, and evolution." *Biology and Philosophy* 16 (5): 595–626.

McGeer, V. 2007. "The regulative dimension of folk psychology." In *Folk Psychology Re-assessed*, 137–156. New York: Springer.

Miller, D. T., and M. Ross. 1975. "Self-serving biases in the attribution of causality: Fact or fiction?" *Psychological Bulletin* 82 (2): 213.

Nisbett, R., and T. Wilson. 1977. "Telling more than we can know: Verbal reports on mental processes." *Psychological Review* 84 (3): 231–259.

Pettigrew, T. F. 1979. "The ultimate attribution error: Extending Allport's cognitive analysis of prejudice." *Personality and Social Psychology Bulletin* 5 (4): 461–476.

Piaget, J. 1969. *The Child's Conception of Physical Causality*. Totawa, NJ: Littlefield, Adams & Co.

Prasada, S., and E. M. Dillingham. 2006. "Principled and statistical connections in common sense conception." *Cognition* 99 (1): 73–112.

Prasada, S., and E. M. Dillingham. 2009. "Representation of principled connections: A window onto the formal aspect of common sense conception." *Cognitive Science* 33 (3): 401–448. doi:10.1111/j.1551-6709.2009.01018.x.

Pronin, E. 2007. "Perception and misperception of bias in human judgment." *Trends in Cognitive Sciences* 11 (1): 37–43.

Pronin, E., D. Y. Lin, and L. Ross. 2002. "The bias blind spot: Perceptions of bias in self versus others." *Personality and Social Psychology Bulletin* 28 (3): 369–381.

Skov, R. B., and S. J. Sherman. 1986. "Information-gathering processes: Diagnosticity, hypothesis-confirmatory strategies, and perceived hypothesis confirmation." *Journal of Experimental Social Psychology* 22 (2): 93–121.

Slowiaczek, L. M., J. Klayman, S. J. Sherman, and R. B. Skov. 1992. "Information selection and use in hypothesis testing: What is a good question, and what is a good answer?" *Memory & Cognition* 20 (4): 392–405.

Smith, E. R. 1990. "Content and process specificity in the effects of prior experiences." In *Advances in Social Cognition*, edited by Thomas K. Srull and Robert S. Wyer, 1–59. Hillsdale, NJ: Erlbaum.

Snyder, M., B. H. Campbell, and E. Preston. 1982. "Testing hypotheses about human nature: Assessing the accuracy of social stereotypes." *Social Cognition* 1 (3): 256–272.

Spaulding, S. 2015. "On direct social perception." *Consciousness and Cognition* 36: 472–482.

Spaulding, S. 2016. "Mind misreading." *Philosohical Issues* 26 (1): 422–440.

Spaulding, S. 2017a. "Do you see what I see? How social differences influence mindreading." *Synthese*.

Spaulding, S. 2017b. "On whether we can see intentions." *Pacific Philosophical Quarterly* 98 (2): 150–170.

Spencer, S. J., C. Logel, and P. G. Davies. 2016. "Stereotype threat." *Annual Review of Psychology* 67: 415–437.

Tetlock, P. E. 1992. "The impact of accountability on judgment and choice: Toward a social contingency model." *Advances in Experimental Social Psychology* 25: 331–376.

Vasilyeva, N., D. Wilkenfeld, and T. Lombrozo. 2017. "Contextual utility affects the perceived quality of explanations." *Psychonomic Bulletin & Review*. doi:10.3758/s13423-017-1275-y.

Vorauer, J. D., A. J. Hunter, K. J. Main, and S. A. Roy. 2000. "Meta-stereotype activation: Evidence from indirect measures for specific evaluative concerns experienced by members of dominant groups in intergroup interaction." *Journal of Personality and Social Psychology* 78 (4): 690.

Westra, E. 2017. "Stereotypes, theory of mind, and the action-prediction hierarchy." *Synthese*: 1–26.

Wilkenfeld, D., J. Gleason, and T. Lombrozo. Forthcoming. "Did he hallucinate because he has schizophrenia?"

Wood, T., and E. Porter. 2018. "The elusive backfire effect: Mass attitudes' steadfast factual adherence." *Political Behavior*. doi:10.1007/s11109-018-9443-y.

Zawidzki, T. W. 2013. *Mindshaping: A New Framework for Understanding Human Social Cognition*. Cambridge, MA: MIT Press.

5 Model theory

The previous chapters discuss phenomena neglected by the most well-developed contemporary theories of mindreading. In this chapter, I evaluate existing theories of mindreading in light of these phenomena. Although the existing theories are good insofar as they go, I argue that we need a theory of mindreading that encompasses the broader set of phenomena I canvass in earlier chapters. I argue that Model Theory has the potential to be the kind of comprehensive, unifying theory of mindreading that we need.

5.1 The function of theories

Theories consist in abstract generalizations about some target phenomenon. Scientific theories' abstract generalizations are empirical and descriptive in nature. Scientific theories have at least two distinct functions. First, scientific theories are supposed to explain the target phenomenon. That is, they ought to provide some empirically testable, unified way of accounting for all the phenomena in the relevant domain.[1] The second function of scientific theories is to make empirically testable predictions about the target phenomena, which are often used for interventions of various sorts. There may be other desiderata for scientific theories as well, but these are the main two.[2] The philosophy of science is littered with debates about how exactly to understand these functions of scientific theories, which is primary, what exactly counts as empirically testable, and whether both explanation and prediction are necessary requirements for every scientific theory. Nothing in my project hangs on how those debates turn out, and I will not presuppose answers to these thorny debates. My claim simply is that scientific theories typically offer empirically testable explanations and predictions

In contrast to scientific theories, the abstract generalizations of philosophical theories typically are not merely empirical and descriptive. Philosophical theories often are *normative*. Utilitarianism, for instance, does not predict that people *will* act to maximize overall happiness. It is a normative theory about the concept of moral goodness and how people *should* act. Similarly, reliabilism does not predict that people will form beliefs only on the basis of truth-conducive belief-forming processes. It claims that beliefs are justified in this way and this is the *best* way for people to form beliefs. Philosophical theories often involve such normative elements. Moreover, philosophical theories can involve different kinds of methodology. Some philosophical methodologies focus on theory building, whereas others focus on comparing, contrasting, and applying theories.[3] There is not just one general methodology for philosophical theories, and there is not one general method for evaluating philosophical theories. In contrast, scientific theories, generally speaking, are evaluated primarily on their empirical testability. Given these key differences between philosophical and scientific theories, we must evaluate them differently.

It is not a universal requirement that philosophical theories accurately describe and make empirically testable predictions about the target phenomena. In many cases of philosophical theorizing, the aim generally speaking is simply to give a coherent and plausible conceptual analysis of some phenomenon. There is no general formula for evaluating coherence and plausibility. Some suggest tallying something like *plausibility points* (Enoch 2011, 267–271). Others argue that philosophical theories aim to achieve an equilibrium of one's reflectively endorsed ideas, and so the evaluation of plausibility hangs on whether the equilibrium positions we have discovered can withstand examination (Lewis 1983; Beebee 2017). However, even these methods for evaluating coherence and plausibility are not decisive. Plausibility judgments are notoriously subjective in philosophy. As they say, one person's *modus ponens* is another person's *modus tollens*. Moreover, all philosophical theories are subject to counterexamples, and the number of counterexamples a given theory faces often has more to do with how prominent the theory is than its plausibility. We can at least say that the theory ought to make no obviously false claims and contain no internal logical inconsistencies, but these are weak constraints. It is quite difficult to formulate general standards of evaluation for philosophical theories.[4] This makes it difficult to see how exactly we should decide between competing philosophical theories.

Theories in philosophy of cognitive science sit at the intersection of philosophical and scientific theories. These theories aim to provide coherent and plausible analyses of empirical phenomena about how the mind works in a way that is empirically tractable. Theories in philosophy of cognitive science often have both normative and descriptive, empirical elements. As such, they are different from both typical philosophical and typical scientific theories.[5] Theories in philosophy of cognitive science may not focus as directly on precise empirical predictions as scientific theories typically do. However, their analyses are meant to accurately describe some empirical phenomenon rather than *just* give a normative analysis, so the empirical claims ought to be empirically testable at least in principle. Thus, how we evaluate theories in philosophy of cognitive science differs from how we evaluate typical philosophical and typical scientific theories.

I articulate the difference between these types of theories because it is rarely clear how we are supposed to evaluate theories in philosophy of cognitive science in general and theories about folk psychology in particular. I propose that these theories ought to provide unified explanations of the target phenomena, that these explanations be conceptually coherent and at least in principle empirically testable, and that they provide a basis for empirical predictions, interventions, and generalizations. Theories in philosophy of cognitive science need not actually be subject to empirical testing to be good and useful (unlike scientific theories), but theorizing ought to be constrained by existing empirical data and testable at least in principle (unlike purely philosophical theories).[6]

5.2 General theories of mindreading

Now that we have some basis for evaluating theories in philosophy of cognitive science, I will apply these evaluative criteria to existing general theories of mindreading. The main contemporary general theories are TT-centric hybrid theories and ST-centric hybrid theories. I will briefly review some of the many varieties of these general theories of mindreading below.

Theory theorists hold that our capacity for mindreading is underwritten by an information-rich body of folk psychological information that we employ to infer others' mental states and predict their behavior. The TT holds that theoretical inferences play a foundational role in developing the capacity for mature mindreading, and they continue to underwrite our ability to explain and predict others' behavior even in adults. And though other sorts of cognitive processes, such as simulation, may play a role in mindreading, these other processes are subsidiary to theoretical inferences. These days, there are many competing accounts of the nature of these theoretical inferences underlying mindreading. Some versions of TT adhere to a strongly modular view of the mindreading capacity (Fodor 1983; Scholl and Gao 2013) according to which mindreading inferences are domain specific and informationally encapsulated, though most contemporary modular versions of TT adopt a weaker notion of modularity (Carruthers 2006; Currie and Sterelny 2000). Some contemporary TT accounts adopt a two-systems approach to mindreading (Apperly and Butterfill 2009) according to which System 1 mindreading processes are early developing, relatively inflexible, and fast, and System 2 mindreading processes are later developing, flexible, and slower. Other theory theorists reject this view in favor of a one-system account (Westra 2017c; Carruthers 2016). Some accounts of TT are nativist (Scholl and Leslie 1999; Carey 2009; Baillargeon 2008), while others are empiricist or constructivist (Wellman 2015; Perner 2010; Heyes 2014). More recently, theory theorists have constructed models of mindreading that are based on Bayesian models of learning and predictive coding (Baker, Saxe, and Tenenbaum 2011; Westra 2017a, d; Gopnik and Wellman 2012). These are just some of the dimensions along which contemporary versions of the TT vary. To make things more complicated, these dimensions cross-cut each other. For instance, one can adopt a Bayesian framework to argue for an empiricist view of mindreading, or one can adopt this framework in support of a nativist view. This variety makes it difficult to make both specific and general claims about contemporary versions of the TT.

In contrast to the TT, simulation theorists argue that simulational processes are developmentally fundamental in the sense that without simulation, we would not have imitation, joint attention, and empathy, which are the building blocks of mature mindreading. Furthermore, ST holds that simulation processes continue to underwrite our developed capacity to explain and predict others' behavior as adults. Hybrid ST theories allow that theoretical processes may supplement simulation, and in some cases theoretical inferences may be more appropriate, but simulational processes are both more foundational and more common than theoretical processes. Contemporary discussions of the ST have not generated the kind of diversity of views that we see with the TT, but the latest versions of the ST have come a long way from the initial versions posited decades ago. For example, now there are versions of two-systems or dual-process approaches for the ST, as well. On a two-systems simulationist approach, System 1 mindreading processes, or *low-level* simulations, are neurally-based mental simulations. According to some views, this kind of simulation is neurally implemented by mirror neurons. The idea is that these neurons fire both when one engages in a certain kind of intentional action, has a certain kind of emotion, or has a certain kind of experience and also when observing another person engage in that intention action, or have that emotion, or have that experience. This is supposed to be the mechanism by which we understand what other people are doing, feeling, and ultimately thinking (Goldman 2009, 2006, Ch 6; Gallese 2009; Oberman and Ramachandran 2007).[7] System 2 mindreading, or *high-level* simulation, is constituted by imagination-based simulation. Unlike low-level simulation, high-level simulation is mediated by psychological concepts, relatively slow, and subject to conscious control (Currie and Ravenscroft 2002; Goldman 2006, Ch 7). Some versions of the ST rely heavily on our ability to introspect our mental simulations. Introspective versions of ST hold that we imagine ourselves in a target's situation, figure out what we would think, feel, and do in that situation, and then attribute that to the target (Goldman 2006, Ch 7), whereas non-introspective versions of ST involve no introspection at all (Gordon 2007). Rather, imagining ourselves in the target's situation constitutes the mental simulation; there is no additional need to introspect. Like contemporary versions of the TT, contemporary versions of the ST vary significantly.

In some ways, the existing general theories of mindreading aim to explain a vast array of phenomena. Not only do they aim to explain the psychological and neural mechanisms underlying mature mindreading, theorists apply the various versions of TT and ST to questions about how mindreading could have evolved (e.g., Schulz 2011), whether and to what extent non-human animals mindread (e.g., Andrews and Beck 2017), children's patterns of success and failure in mindreading tasks (e.g., Wellman, Fang, and Peterson 2011), the ontogeny of mindreading in relation to language, executive function, and hypothetical reasoning (e.g., Breheny 2006), the idiosyncratic patterns of social inference of people on the Autism Spectrum Disorder (Anderson and Cushing 2012, e.g.,), etc.[8] In their foundational work on hybrid theories, Nichols and Stich even lay out a broad checklist of facts that theories of mindreading ought to explain which includes many of the topics mentioned above (Nichols and Stich 2003, 77). This is a broad and varied list of *explananda* by any measure.

However, in other ways, existing general theories of mindreading take a myopic view of what needs to be explained. General theories of mindreading, and even theories in specific debates about mindreading, tend to assume that the only mode of mindreading (worth discussing) is accurately attributing mental states in order to accurately explain and predict a target's behavior.[9] This is probably due at least in part to the influence of the false-belief task on subsequent theorizing about mindreading. I have argued over the course of the last few chapters of this book that accurately explaining and predicting behavior is just one small part of our mindreading abilities. The way in which we socially categorize people, whether we perceive others to be relevantly similar to us, the biases we bring to bear on a situation, the effects of situational context, our goals in a social interaction, the mindreading strategies we adopt, and the kind of explanation mindreading produces all can vary. This makes for an incredibly diverse set of mindreading practices. Importantly, these various phenomena reflect diversity in mindreading itself, not just how we use subsequently use mindreading. Most contemporary versions of the TT and the ST do not discuss these phenomena even though they are directly relevant to understanding the processes of mindreading.

What does this mean for how general theories of mindreading fare with respect to the evaluative criteria articulated in the previous section? Given the restricted phenomena that traditional general mindreading theories consider, the various versions of TT and ST do a decent job of explaining mindreading phenomena. Both kinds of theories describe a set of cognitive processes or mechanisms, broadly construed, that underwrite our ability to accurately attribute mental states to others in order to explain and predict their behavior.

To take just one example, consider Goldman's dual-process version of the ST. This account aims to explain the ontogenetic development of accurately attributing mental states to other agents; the psychological implementation of high-level simulation-based accurate attributions of beliefs and desires; the neural implementation of low-level resonance-based accurate inferences about emotions, intentions, and sensation; how simulation underlies emotional contagion and empathy (discussion of both are limited to cases of *accurate* contagion and empathy); and other topics involving accurate attributions of mental states. Indeed, the focus on accurate mindreading is clear in Goldman's list of "central questions in the field" (2006, 21).[10]

The only places in which inaccurate or distorted mindreading come up in Goldman's book are in discussing the idiosyncratic social inferences of people with Autism Spectrum Disorder (2006, 200–206) and data on introspective confabulation (2006, 234–238). Nowhere does Goldman discuss neurotypical adults' distorted or inaccurate attributions of mental states to others, the various psychological processes that underlie these other modes of mindreading, the conditions that bring about these other mindreading processes, or how our goals or perceived similarity to a target influence the process and accuracy of mindreading others. This is not because Goldman's project has a narrow, restricted focus. He aims to offer an impressively comprehensive theory of mindreading.

Goldman notes in the first chapter of his book that the subject of mindreading is complex because so many disciplines study the phenomenon. "Serious students of the field cannot ignore any of these disciplines, because each contributes valuable insights" (Goldman 2006, 4). I agree wholeheartedly. The problem is that Goldman and *many* others have overlooked important, relevant research on mindreading from social psychology. When you take into consideration all the mindreading phenomena, it is clear that most general theories of mindreading do a poor job of unifying phenomena and have very little inductive utility for the *full range* of mindreading. These theories typically do not even discuss the phenomena described in the previous two chapters; thus, they offer no *unified* or *coherent* explanation of these phenomena. They provide no basis for predicting how one's own social identity and the social identity of the target(s) will affect one's mindreading, or how one's goals will affect both the mindreading process and product, or how these processes break down. This makes many of the existing general theories of mindreading empirically inadequate according to the evaluative criteria I articulate above.

Indeed, given the complexity and interactions amongst goals, approaches, and social categorization reviewed so far, one would expect that the predictive capacity of theory theories and simulation theories would be relatively poor even for the restricted set of phenomena they consider. Even when we do aim for accuracy, so many factors influence and interfere with accurate mindreading.[11] Given the assumptions of existing mindreading theories, they either will make false predictions about such cases or simply dismiss them as noise. Neither option is very appealing. False predictions clearly are problematic. Furthermore, these cases are indicative of a greater deal of complexity and diversity in mindreading than standard mindreading theories recognize. That is, they are not simply noise.

5.3 Model theory

In this section, I shall argue that Model Theory has the potential to be a comprehensive, unified, and plausible account of mindreading. Model Theory was originally proposed by Heidi Maibom (2007, 2009, 2003) as a more promising hybrid version of the TT. Peter Godfrey-Smith (2005) developed a slightly different version of Model Theory that was built upon his prior work on the use of models in science. I shall explain and elaborate this theory.

In general, Model Theory conceives of mindreading as a form of theoretical modeling. As such, Model Theory is a version of the TT.[12] The inspiration for the view comes from a view in philosophy of science that rejects the idea that theories are systems of axiomatized sentences. The alternative view holds that theories are models or families of models, and scientific activity consists of constructing and applying models. On this view, models are the central unit of scientific theorizing. There are more pluralistic views out there, which hold that modeling is just one kind of scientific activity amongst others, and that sometimes applying a model or family of models is appropriate whereas other times using theories as they are traditionally conceived is appropriate (Frigg and Hartmann 2006). Though pluralism is often wise, we need not take a stand on that issue in order to use the Model Theory of mindreading.

Model Theory holds that mindreading consists in deploying a model psychological profile of a target. The kind of models posited by Model Theory are conceptual models, which involve systems of related concepts.[13] Models are hypothetical representations that specify a general structure, relations, and properties of some phenomenon. These hypothetical representations can be used to represent more complex phenomena. For mindreading, we construct and apply simplified model psychological profiles of targets in order to understand complex social interactions. There is a basic folk psychological model, which consists in a distinction between beliefs and desires, the idea of sensory input and behavioral output, and characteristic dependence of action on perceptions, memories, goals, and temptations (Godfrey-Smith 2005, 10).[14] Elements of this core folk psychological model are innately specified, which explains why basic aspects of folk psychology appear to be common across cultures (Malle 2008).

The core folk psychological model can be elaborated in various ways and is the basis for many culturally and even individually specific models. If this is right, it would go some way toward explaining how there are both universal features of folk psychology and cultural and individual variation as well (e.g., Liu et al. 2008; Sabbagh et al. 2006; Norenzayan, Choi, and Nisbett 2002). We may construct specific model psychological profiles for different social groups and perhaps even specific individuals. Over time and with experience, we come to incorporate into our folk psychological models knowledge of different kinds of social norms, institutions, and social roles, individualized knowledge about a particular person's history and personality, knowledge of stereotypes, social biases, etc. Elements can be introduced to a folk psychological model of a target via explicit teaching, implicit learning, automatic cognitive processing, from one's own experiences, etc. Once an element is introduced to the model, it will acquire connections to a range of other related elements, thereby becoming part of a richer structure (Godfrey-Smith 2005, 8–9).

The mindreading models we employ can be more or less elaborate, with some being mere schema we generate on the fly and others being detailed representations of individuals and social groups. Folk psychological models may be explicit and deployed deliberately, such as in scientific interpretation, but they need not be. The models may be implicit, and the agent using them may not be able to describe them in any great detail, and thus may not be able to articulate the similarities between a model and the target phenomenon. In such a case, the result of the interpretation is that the agent simply sees the target phenomenon as an instance of a relatively familiar model.

So far, I have described what folk psychological models are and emphasized their diversity. There is also variety in how we employ these folk psychological models. We put these models to different uses depending on our interests and the context. We can apply detailed models carefully, checking to see if the model matches the behavior of the target. One primary factor in determining model use is whether something important hangs on mindreading correctly, if it matters to you personally, or if the situation is highly unusual. In these cases, you will tend to use a more detailed model in order to generate more accurate model of the other person's mind. You may use this model to explain a target's behavior (these explanations may be teleological, formal, or mechanistic), make predictions, manipulate the target's behavior, or all of the above, depending on your interests in the situation.

We can also apply mere schematic models in a quick and simple way. In cases where efficiency matters more to you than accuracy, you will use more schematic models in order to have a close enough model without having to deliberate very carefully. The schematic models you use depend on whether the person you are mindreading is perceived to be part of an in-group or an out-group. In the former case, you will tend to use a model roughly based on what you take yourself to think, feel, and do in various situations. In the latter case, your model will be based on relevant stereotypes. You may use these schematic models to categorize the target (e.g., as characteristic of the relevant social stereotype or as like me), explain behavior (e.g., formal or teleological explanations), or make predictions, depending on your interests.

When your motives are more self-serving, what varies is not so much the level of elaboration of the model but how you use the model to cement preconceived ideas about oneself, one's in-group, and various out-groups. In these cases, the elements of the model heavily emphasize patterns of behavior in line with your existing values and beliefs. Typically, in such cases, the model is used to explain a target's behavior – in order to justify, rationalize, condemn, or dismiss it – rather than predict or manipulate the target's behavior.

As I emphasized in previous chapters, we may have multiple goals that change in priority throughout a social interaction, and the approaches to mindreading we take may shift accordingly. Model Theory has an elegant explanation of how this can happen. Model Theory distinguishes the models from how we use the models. The models themselves exhibit a great deal of variety based on which elements are introduced how they are introduced. One nice way of understanding Chapter 3's discussion of social categorization, stereotypes, social biases, and situational context is as a survey of how models are elaborated. On top of the variety in model construction, the models can be employed differently as well. In other words, the very same model may be used to categorize behavior, confirm a stereotype, or make rough-and-ready predictions about what a target will do. Moreover, we can change our focus to a different model of a target if our goals or understanding of the situation change more dramatically. One can understand Chapter 4's discussion of the goals, processes, and products of mindreading as a survey of how models are employed. Thus, there is no mystery in how our approaches to social interactions shift along with our goals and interests. We can use models in various ways and shift to different more appropriate models if the situation calls for it.

As I mentioned above, the Model Theory is a version of the TT, but it has an advantage over the ST and many other versions of the TT. In particular, it is well poised to explain the diversity in modes of mindreading and the psychological processes that underlie these various modes, but it also explains the diversity of input to mindreading, the goals that mindreading serves, the different kinds of products of mindreading, and the conditions under which mindreading is likely to be accurate. Most existing hybrid theories are in principle capable of explaining *some* of this diversity, but in practice they only address a very limited range of mindreading processes, and it is an open question whether and how we can alter various existing theories of mindreading to explain all these various features of mindreading. Thus, a distinctive benefit of the Model Theory is that it already explains the diversity of mindreading.

A further advantage is that it is compatible with a number of different views on the ontogenetic development of mature mindreading.[15] I take this to be a theoretical virtue because the developmental story is likely to be highly complex, and despite a wealth of research and exciting recent developments, theorists are still no closer to a consensus on the right ontogenetic account of mindreading. To illustrate just some of the outstanding disagreements, theorists still do not agree on whether infants track *beliefs* or some other type of representation (e.g., Apperly and Butterfill 2009; Carruthers 2013), the folk psychological practices that are language dependent (e.g., Brooks and Meltzoff 2015; San Juan and Astington 2017), whether the standard false-belief task and the non-linguistic false-belief tasks are measuring what they purport to be measuring (e.g., Heyes 2014; Baillargeon, Scott, and He 2010), and which social cognitive capacities are innate and which are learned (e.g., Wellman 2015; Malle 2008; Westra 2017b). Given this state of the field, it is wise for a general theory of mindreading to wait and see how the field turns out. Thus, I take it to be a virtue of the Model Theory that it is compatible with a number of ways the development of mindreading could be realized.

Model Theory also has the virtue of being compatible with evidence of universal elements of folk psychology and evidence of the cultural variability of mindreading practices (Godfrey-Smith 2005, 9).[16] On the view espoused here, elements of the core folk psychological model are universal and innately specified, but the elaborations of this model and the ways in which we apply models are strongly influenced by cultural norms.

An advantage of Model Theory that gets little attention in the mindreading literature is that it takes practice to learn how to apply to models, i.e., to recognize patterns in behavior and apply appropriate models to the phenomena (Maibom 2009). Various versions of the TT and the ST pay lip service to the idea that mindreading requires practice, but the theories do not really capture why and how practice is required. After all, if mindreading just is the application of a theory, your theory will tell you how to make sense of various behaviors. There will be ambiguous cases, but in general once you master the *theory* there is no reason to think learning to theorize will take lots of practice. And if mindreading just is simulational, there may be a bit of practice involved in figuring out what one would do in a target's situation, but really this is quite minimal. In contrast, for Model Theory there are lots of different models that could be appropriate for a situation and that could be employed in different ways. Facility with folk psychological models requires practice constructing and employing lots of different models in lots of different situations. You have to learn which models are appropriate in which situational contexts and how to employ them effectively. This folk psychological development occurs throughout childhood and adulthood (Hughes and Leekam 2004). Often the focus in the mindreading literature is on what happens from age 1–5 and little attention is paid to developments in mindreading after the preschool years. But clearly our mindreading abilities develop quite a lot for many years after that (Bosco, Gabbatore, and Tirassa 2014; Dumontheil, Apperly, and Blakemore 2010), and the Model Theory offers a nice account of what that development involves.

Finally, Model Theory may actually be compatible with some existing theories of mindreading – most likely other versions of the TT. For example, it seems plausible that we could explain how we learn to construct and employ various mindreading models with a Bayesian or predictive coding model of mindreading. It may take significant work to actually make these theories cohere well in practice, but I take it to be good in principle because this could expand the explanatory scope of the theories. For instance, Evan Westra (2017a, d) offers a predictive coding account of mindreading that I think is a promising candidate for combining with Model Theory. Thus, I am not so concerned with establishing that Model Theory is incompatible with every other existing theory of mindreading as I am with constructing an empirically plausible, philosophically sound general account of how we understand and interact with others.

With respect to the evaluative criteria laid out in Section 5.1, the Model Theory does well. It is empirically based in the sense that it is well grounded in the relevant empirical findings, particularly cognitive and social psychology. It is also grounded in a philosophical view of models, based on work in the philosophy of science. It is a coherent and relatively plausible general theory of mindreading.

5.4 Assessment

The Model Theory seems to be a more comprehensive, cohesive, and plausible account of mindreading than many of the existing theories that dominate the mindreading literature. In the last part of this chapter, I will consider a few foreseeable objections to the account I have advanced here. First, one might object that the Model Theory is too noncommittal. After all, it is compatible with many different developmental accounts of mindreading, both universality and cultural dependence of mindreading, and it does not allow us to make predictions about exactly what kind of model an individual will construct or how she will employ that model.

Though Model Theory is broad, it is committed to plenty of specific claims and generates many testable predictions. For example, Model Theory is committed to the idea that mindreading input, processing, and output are much more diverse than most versions of the TT or ST posit. As a result, Model Theory predicts that in real-life cases (as opposed to mindreading tasks in the lab, such as the false-belief task) there will be much more mindreading disagreement because individuals will construct and employ models differently. This is a testable prediction. The existing TT and ST hybrids do not predict such mindreading disagreements amongst competent, mature mindreaders. Our everyday experiences give *prima facie* reason to expect Model Theory's prediction to be borne out. After all, we are all familiar with the frustrating debates about whether a given police officer felt he was in mortal danger, what the police officer knew, or what someone killed by police was intending to do.[17]

Model Theory is compatible with both universality and cultural dependence of mindreading, but it does make specific commitments about both universality and cultural dependence. It predicts that most neurotypical humans will share some basic ways of making sense of intentional behavior, e.g., a distinction between mental representations about how the world is and mental representations about how an agent wants the world to be, the idea of sensory input and behavioral output, the dependence of action on perceptions, memories, goals, and temptations, etc. We expect most people to make sense of others' behavior in these types of terms. The view does not say when in development this core folk psychological model is deployable or how it develops in ontogeny. But, as I argue above, such abstention is wise, because the developmental story is still quite uncertain. Model Theory is committed to the idea that as our mindreading capacities mature, we become capable of relying on more specific, culturally-dependent elaborations of the core folk psychological model. It predicts that individuals with different social experiences will introduce different elements into their models, which inevitably will result in different mindreading judgments.

Model Theory cannot tell us exactly what model an individual will construct and how she will apply it. That depends too much on factors that we cannot decisively determine from our limited knowledge of each individual's experiences, cognitive styles, situational contexts, and goals. We should not regard this limitation as a problem for Model Theory. It is true that there is a lot of diversity in how individuals mindread. However, the Model Theory does give parameters for what kinds of models we will construct in various situations, how our goals influence how we apply these models, and the kind of results we are likely to get. We can also, based on the material from Chapter 4, make predictions about when such modeling is likely to lead to error. Thus, although the theory does not predict exactly how an individual will understand a social interaction, it does make predictions about variables that determine the kind of model we construct and how we will employ it.

In sum, Model Theory makes plenty of empirically based commitments and testable predictions. If many of these predictions turn out to be false, we will have reason to reject the theory. In many cases however, there is strong *prima facie* reason to expect these predictions to turn out to be true.

Finally, one may object that Model Theory is not uniquely suited to be *the* comprehensive theory of mindreading we need. Other theories may be just as well suited to cover the ground discussed in this book if given the chance to expand. To this objection I say *Sure! The more the merrier!* I do not hold that Model Theory is alone in its ability to explain the broad class of mindreading phenomena discussed here. One may be able to overhaul various versions of the TT or ST to make them appropriately comprehensive theories, or one may come up with a completely different general theory of mindreading. I suggested above that one could even combine Model Theory with Bayesian or predictive coding models of mindreading. I think Model Theory is a particularly promising theory, but my main concern in this book is to establish that the mindreading literature has been too myopically focused on one relatively limited mode of mindreading and that mindreading is much more interesting and complex than the current literature recognizes. I encourage others to work on developing Model Theory or other genuinely comprehensive theories of mindreading.

Notes

1 The unified explanation may consist in axiomatized sets of sentences that describe the target phenomena or some sort of model or set of models that represent the target phenomena. These models may be quantitative, e.g., models that mathematically represent the target phenomenon, or the models may be conceptual. Conceptual models come in many varieties, e.g., physical scale models, simulations, or simply a system of related concepts. Whatever form the explanations take, scientific explanations ought to be empirically testable.

2 Simplicity, unity, fruitfulness, and other epistemic values may be legitimate criteria for evaluating scientific theories. Famously, Kuhn argues that these epistemic values are individually imprecise, pull in different directions, and weighted differently by different scientific paradigms (Kuhn 2012). Nevertheless, scientists

often do evaluate scientific theories in light of these criteria, even if none of them is decisive in theory choice.

3 Thanks to Nate Sheff for pointing out this distinction.

4 It is plausible that there are other evaluative criteria that are used in some subfields philosophy and for some philosophical methodologies. Marcus Arvan (2016) argues that there are seven principles of theory selection at use in moral philosophy: firm foundations, internal coherence, external coherence, explanatory power, unity, parsimony, and fruitfulness. These evaluative criteria certainly are used in moral philosophy and other areas, but it is unlikely that each of them is appropriate in every subfield and every methodology in philosophy. Parsimony, for example, is often a criterion of evaluation in metaphysical debates about theory construction, though it may be less appropriate in other subfields like normative ethics or with other methodologies such as history of philosophy. See Huemer (2009), Machery (2017), and Sober (2015) for more on parsimony in philosophical debates.

5 Philosophy of cognitive science is not unique in this way. Other interdisciplinary subfields have similar characteristics, e.g., philosophy of biology, biomedical ethics, philosophy of education, etc.

6 I do not regard these as controversial evaluation criteria, and certainly there are good theories in philosophy of cognitive science that do not meet some of these conditions. The point is to have some way to evaluate interdisciplinary theories, not to lay out demarcation criteria.

7 But see Spaulding (2012, 2013) for a critique of these claims.

8 Certainly, one can take part in these more specific debates without adhering to any general theory about mindreading. I focus on general theories of mindreading here because, unlike views in the specific debates described above, they aim to provide comprehensive views of our mindreading abilities and guide our theorizing about these more specific debates. As I shall argue shortly, these general theories of mindreading fail to be comprehensive. As such, the overall views of mindreading and their application to specific debates are misguided.

9 That is, mindreading theories aim to explain how this capacity to accurately attribute mental states to explain and predict behavior evolved, how young children develop this capacity, what underwrites adults' capacity to accurately attribute mental states, etc.

10 These central questions, paraphrased, are: (1) How do people attribute mental states to others? (2) How do people mindread themselves? (3) How is the mindreading capacity acquired? (4) What are the contents of people's concepts of mental states? (5) How is mindreading related to other cognitive processes in other domains? (6) What is the relationship between mindreading and other forms of social cognition, like empathy and imitation? (7) How did mindreading evolve? See Goldman (2006, 21).

11 See Chapter 4 for more on how the various approaches to mindreading lead to error.

12 It is actually quite tricky to distinguish simulational models from theoretical models because simulations can take many forms. See Spaulding (2012) for more on what should count as evidence for simulational modeling. Simulational models operate either on the principle of *reuse* or *resemblance*. The models posited by Model Theory are not necessarily based on reusing one's cognitive mechanisms offline to simulate a target's cognitive processes, nor do the operation of the models necessarily resemble the cognitive processes of the target.

The models *may* resemble the target's cognitive processes (and in that sense, the view allows simulational modeling), but this is not a requirement or generally how the models work (and, thus, in that sense the view is a hybrid TT).

13 See Section 4.1 above and Frigg and Hartmann (2006) for more on different conceptions of models.

14 Maibom (2007, 2009) distinguishes three types of models relevant to social interpretation: models for goal-directed behavior, social models, and folk psychological models. In the main text, I will simply focus on the core folk psychological model and specific elaborations of this model for individuals and social groups. Depending on what one aims to explain, these distinctions may make sense. However, I do not think it is helpful to treat them as distinct in constructing an account of mindreading. As I argue in previous chapters, interpreting goal-directed behavior, deploying information about social norms and schemata, and interpreting others' mental states are messily intertwined cognitive processes in mindreading. Understanding real-life mindreading practices requires appreciating how these and other processes interact. This does not reflect an objection to Maibom's taxonomy so much as a difference in focus. She is focused on understanding the varieties of theoretical structures we use to understand the social world, while I am concerned with capturing the full range of mindreading phenomena.

15 This advantage is not *unique* to Model Theory, of course.

16 Though I do not discuss this possibility in the book, folk psychological models may be applied to non-humans as well, e.g., animals, nations, corporations, etc.

17 I discuss the Eric Garner case at length in Spaulding (2017a) and the Tamar Rice case in Spaulding (2017b).

References

Anderson, J. L., and S. Cushing. 2012. *The Philosophy of Autism*. Lanham, MD: Rowman & Littlefield Publishers.

Andrews, K., and J. Beck. 2017. *The Routledge Handbook of Philosophy of Animal Minds*. New York: Routledge.

Apperly, I. A., and S. A. Butterfill. 2009. "Do humans have two systems to track beliefs and belief-like states?" *Psychological Review* 116 (4): 953.

Arvan, M. 2016. *Rightness as Fairness: A Moral and Political Theory*. New York: Springer.

Baillargeon, R. 2008. "Innate ideas revisited: For a principle of persistence in infants' physical reasoning." *Perspectives on Psychological Science* 3 (1): 2–13.

Baillargeon, R., R. Scott, and Z. He. 2010. "False-belief understanding in infants." *Trends in Cognitive Sciences* 14 (3): 110–118.

Baker, C., R. Saxe, and J. Tenenbaum. 2011. "Bayesian theory of mind: Modeling joint belief-desire attribution." Proceedings of the Annual Meeting of the Cognitive Science Society.

Beebee, H. 2017. "Philosophical scepticism." *Proceedings of the Aristotelian Society*.

Bosco, F. M., I. Gabbatore, and M. Tirassa. 2014. "A broad assessment of theory of mind in adolescence: The complexity of mindreading." *Consciousness and Cognition* 24: 84–97.

Breheny, R. 2006. "Communication and folk psychology." *Mind and Language* 21 (1): 74–107.

Brooks, R., and A. N. Meltzoff. 2015. "Connecting the dots from infancy to childhood: A longitudinal study connecting gaze following, language, and explicit theory of mind." *Journal of Experimental Child Psychology* 130: 67–78.

Carey, S. 2009. *The Origin of Concepts*. Oxford: Oxford University Press.

Carruthers, P. 2006. *The Architecture of the Mind*. Oxford: Oxford University Press.

Carruthers, P. 2013. "Mindreading in infancy." *Mind and Language* 28 (2): 141–172.

Carruthers, P. 2016. "Two systems for mindreading?" *Review of Philosophy and Psychology* 7 (1): 141–162.

Currie, G., and I. Ravenscroft. 2002. *Recreative Minds: Imagination in Philosophy and Psychology*. Oxford: Oxford University Press.

Currie, G., and K. Sterelny. 2000. "How to think about the modularity of mindreading." *The Philosophical Quarterly* 50 (199): 145–160.

Dumontheil, I., I. A. Apperly, and S.-J. Blakemore. 2010. "Online usage of theory of mind continues to develop in late adolescence." *Developmental Science* 13 (2): 331–338.

Enoch, D. 2011. *Taking Morality Seriously: A Defense of Robust Realism*. Oxford: Oxford University Press on Demand.

Fodor, J. A. 1983. *The Modularity of Mind*. Cambridge, MA: MIT Press.

Frigg, R., and S. Hartmann. 2006. "Models in science." In *Stanford Encyclopedia of Philosophy*. Stanford: Stanford University Press.

Gallese, V. 2009. "Mirror neurons, embodied simulation, and the neural basis of social identification." *Psychoanalytic Dialogues* 19 (5): 519–536.

Godfrey-Smith, P. 2005. "Folk psychology as a model." *Philosophers' Imprint* 5 (6): 1–16.

Goldman, A. I. 2006. *Simulating Minds: The Philosophy, Psychology, and Neuroscience of Mindreading*. New York: Oxford University Press.

Goldman, A. I. 2009. "Mirroring, simulating and mindreading." *Mind & Language* 24 (2): 235–252.

Gopnik, A., and H. M. Wellman. 2012. "Reconstructing constructivism: Causal models, Bayesian learning mechanisms, and the theory theory." *Psychological Bulletin* 138 (6): 1085.

Gordon, R. M. 2007. "Ascent routines for propositional attitudes." *Synthese* 159 (2): 151–165.

Heyes, C. 2014. "False belief in infancy: A fresh look." *Developmental Science* 17 (5): 647–659. doi:10.1111/desc.12148.

Huemer, M. 2009. "When is parsimony a virtue?" *The Philosophical Quarterly* 59 (235): 216–236. doi:10.1111/j.1467-9213.2008.569.x.

Hughes, C., and S. Leekam. 2004. "What are the links between theory of mind and social relations? Review, reflections and new directions for studies of typical and atypical development." *Social Development* 13 (4): 590–619.

Kuhn, T. S. 2012. *The Structure of Scientific Revolutions*. Chicago, IL: University of Chicago Press.

Lewis, D. 1983. *Philosophical Papers Vol. I*. Vol. 1. Oxford: Oxford University Press.

Liu, D., H. M. Wellman, T. Tardif, and M. A. Sabbagh. 2008. "Theory of mind development in Chinese children: A meta-analysis of false-belief understanding across cultures and languages." *Developmental Psychology* 44 (2): 523.

Machery, E. 2017. *Philosophy Within Its Proper Bounds*. Oxford: Oxford University Press.

Maibom, H. 2003. "The mindreader and the scientist." *Mind and Language* 18 (3): 296–315.

Maibom, H. 2007. "Social systems." *Philosophical Psychology* 20 (5): 557.

Maibom, H. 2009. "In defence of (model) theory theory." *Journal of Consciousness Studies*, 16 (6–8): 360–378.

Malle, B. F. 2008. "The fundamental tools, and possibly universals, of human social cognition." In *Handbook of Motivation and Cognition Across Cultures*, 267–296. San Diego, CA: Academic Press.

Nichols, S., and S. Stich. 2003. *Mindreading: An Integrated Account of Pretence, Self-Awareness, and Understanding Other Minds, Oxford Cognitive Science Series*. Oxford: Oxford University Press.

Norenzayan, A., I. Choi, and R. Nisbett. 2002. "Cultural similarities and differences in social inference: Evidence from behavioral predictions and lay theories of behavior." *Personality and Social Psychology Bulletin* 28 (1): 109–120.

Oberman, L. M., and V. Ramachandran. 2007. "The simulating social mind: The role of the mirror neuron system and simulation in the social and communicative deficits of autism spectrum disorders." *Psychological Bulletin* 133 (2): 310–327.

Perner, J. 2010. "Who took the cog out of cognitive science?" In *Cognition and Neuropsychology: International Perspectives on Psychological Science*, edited by R. Schwarzer and P. A. Frensch, 241–262. New York, NY: Psychology Press.

Sabbagh, M. A., F. Xu, S. M. Carlson, L. J. Moses, and K. Lee. 2006. "The development of executive functioning and theory of mind: A comparison of chinese and U.S. Preschoolers." *Psychological Science* 17 (1): 74–81. doi:10.1111/j.1467-9280.2005.01667.x.

San Juan, V., and J. W. Astington. 2017. "Does language matter for implicit theory of mind? The effects of epistemic verb training on implicit and explicit false-belief understanding." *Cognitive Development* 41: 19–32.

Scholl, B. J., and T. Gao. 2013. "Perceiving animacy and intentionality: Visual processing or higher-level judgment." *Social Perception: Detection and Interpretation of Animacy, Agency, and Intention* 4629.

Scholl, B. J., and A. M. Leslie. 1999. "Modularity, development and 'theory of mind.'" *Mind & Language* 14 (1): 131–153.

Schulz, A. W. 2011. "Simulation, simplicity, and selection: An evolutionary perspective on high-level mindreading." *Philosophical Studies* 152 (2): 271–285.

Sober, E. 2015. *Ockham's Razors*. Cambridge: Cambridge University Press.

Spaulding, S. 2012. "Mirror neurons are not evidence for the Simulation Theory." *Synthese* 189 (3): 515–534.

Spaulding, S. 2013. "Mirror neurons and social cognition." *Mind & Language* 28 (2).

Spaulding, S. 2017a. "Do you see what I see? How social differences influence mindreading." *Synthese*.

Spaulding, S. 2017b. "How we think and act together." *Philosophical Psychology* 30 (3): 302–318.

Wellman, H. M. 2015. *Making Minds: How Theory of Mind Develops*. Oxford: Oxford University Press.

Wellman, H. M., F. Fang, and C. C. Peterson. 2011. "Sequential progressions in a theory-of-mind scale: Longitudinal perspectives." *Child Development* 82 (3): 780–792.

Westra, E. 2017a. "Character and theory of mind: An integrative approach." *Philosophical Studies*.

Westra, E. 2017b. "Pragmatic development and the false belief task." *Review of Philosophy and Psychology* 8 (2): 235–257.

Westra, E. 2017c. "Spontaneous mindreading: A problem for the two-systems account." *Synthese* 194 (11): 4559–4581.

Westra, E. 2017d. "Stereotypes, theory of mind, and the action-prediction hierarchy." *Synthese*: 1–26.

6 Epistemic and ethical applications

The comprehensive view of mindreading I articulate and defend in the previous chapters has implications far beyond philosophy of psychology. In this chapter, I take the broader conception of mindreading advanced in this book and apply it to debates in epistemology and ethics. In particular, I argue that this view of mindreading can shed light on how we judge others' knowledge and competence, and this has direct implications for debates about the epistemology of peer disagreement and various interventions on epistemic injustice.

6.1 The isolation of the mindreading literature

Most philosophical discussions of mindreading stay squarely within the realm of philosophy of psychology. Theorizing about mindreading plays a role in debates about the modularity of the mind, the representational theory of mind, language development, the semantics of ordinary language use, etc. Using mindreading as a case study for understanding the mind and our capacity for language makes sense. There is a wealth of philosophical and empirical work on the topic, it is an excellent example of fruitful interdisciplinary interaction and collaboration, mindreading is ubiquitous in our everyday lives, and the capacity for mature mindreading seems to be a distinctively human trait. Thus, we can see why the study of mindreading would have natural applications in other areas of philosophy of psychology.

One would think that understanding how we interpret and interact with others is important for other philosophical areas, as well. Various topics in epistemology and ethics, for example, concern how we know what others believe, our judgments that others are knowledgeable and competent with respect to some issue, whether we regard someone as an epistemic inferior, peer, or superior, and all the moral judgments entangled with these inferences. These inferences clearly involve mindreading, so it is initially puzzling that the mindreading literature has had so little to say about these epistemic and ethical topics. Mindreading theorists have not engaged with these literatures, and ethicists and epistemologists have not recognized the relevance of mindreading to these debates.

One of the main reasons for this lack of interaction is that mindreading has been so narrowly construed by mindreading theorists. As I discuss in the previous chapters, most theorizing about mindreading posits one mode of mindreading: correctly inferring mental states to accurately explain and predict a target's behavior. This mode of mindreading is somewhat relevant to epistemic and ethical judgments, but this narrow construal makes it difficult to see just how interesting and relevant mindreading is for these epistemic and ethical debates. As we know now, mindreading is much more diverse than the narrow construal of it suggests. There is variety in what we take as input to mindreading, our approaches to mindreading, and what kind of judgments we end up producing. People with very different social experiences, judgments of perceived similarity, and goals will socially categorize differently, employ different social biases, approach social inferences differently, and, in the end, produce very different social interpretations. Understood in this way, mindreading is much more complex, epistemically interesting, and ethically murky.

In the rest of this chapter, I shall apply the more comprehensive view of mindreading developed in this book to debates in social epistemology and ethics. The theory of mindreading I articulate in the previous chapter can help shed light on how we judge who is an epistemic peer, analyze disagreements amongst people we judge to be epistemic peers, give us guidance on when and how these judgments are biased and on epistemic injustice, provide perspective on interventions on epistemic injustice, and give guidance on navigating contentious discussions.

6.2 Judging others' knowledge and competence

Judgments about others' knowledge and competence play a key role in debates about peer disagreement and epistemic injustice. The epistemology of peer disagreement concerns what we ought to do when someone we take to be an epistemic peer disagrees with us. Two people are epistemic peers with respect to X when they possess the same evidence about X and are equally intelligent, free from bias, and competent at perceiving, reasoning, etc. (Kelly 2011). Implicit in this literature is the notion that typically we are astute judges of who is an epistemic inferior, epistemic peer, and epistemic superior. Judging others' knowledge and competence plays a key role in the discussion of epistemic injustice, as well. Epistemic injustice, in particular testimonial injustice, occurs when a hearer's prejudices result in downgrading a speaker's credibility (Fricker 2007). That is, in virtue of epistemically irrelevant facts about the speaker, the hearer downgrades the speaker's epistemic status. Central to both philosophical debates is the issue of how we judge others' knowledge, intelligence, reasoning abilities, biases, etc. The mindreading phenomena canvassed in this book illuminate how we make these judgments about others' knowledge and competence and when to doubt our judgments.

How do we decide when someone is our epistemic peer, inferior, or superior about some issue? Specifically, how do we decide that another person has comparable, less, or more knowledge and competence with respect to some issue? The first thing to note is that judgments about others' knowledge and competence often are comparative. We judge how knowledgeable and competent a target is in comparison to *ourselves*. However, one factor that impedes objective comparative judgments is that people tend to overestimate their own knowledge and competence.

As I discuss in Chapter 4, we are subject to several very common self-enhancing biases that skew our judgments about our own and other people's knowledge and competence. For example, the Self-Serving Attributional bias describes our tendency to take credit for our success and deny responsibility for our failure (Miller and Ross 1975). We often attribute our successes to some internal factor, e.g., diligence, talent, knowledge, and attribute our failures to external mitigating factors, e.g., bad luck or bias. In this way, we come to overestimate our own competence and underestimate others' competence. Thus, even when someone is in fact an epistemic peer, we may not regard her as such because of this self-enhancing bias.

Relatedly, the Dunning-Kruger effect is a cognitive bias wherein poor performers in social and intellectual domains are unaware of their incompetence (Kruger and Dunning 1999). I did not discuss this bias in the previous chapters because it regards self-evaluation more than other-evaluation, but it is relevant here. The Dunning-Kruger effect describes the tendency of individuals who are unknowledgeable or incompetent with respect to some issue to egregiously overestimate their own knowledge and competence, and fail to recognize others' equal or superior knowledge and competence. The deficiency of the comparatively ignorant and incompetent is invisible to them presumably because recognizing their deficiency requires the very competency they lack.

Another self-enhancing bias relevant to judging epistemic peers is Naïve Realism, which I describe in Chapter 4. Naïve Realism is the tendency to regard others as more susceptible to bias and misperception than oneself (Pronin, Lin, and Ross 2002; Pronin 2007). We think we simply see things as they are, but others suffer from bias. We regard ourselves and others in our relevant in-groups as perceiving the world as it truly is, but we regard those who disagree with us as misguided, misinterpreting, or biased by their personal motivations. This bias is particularly relevant to the epistemology of peer disagreement. The empirical data suggest that simply in virtue of the fact that someone disagrees with us, we will downgrade that person's epistemic status in relation to our own (Pronin, Lin, and Ross 2002, 378–379; Kennedy and Pronin 2008).

So far, I have discussed three self-enhancing biases that influence how we judge our own knowledge and competence in relation to others. This is only half the story, though. The other half concerns how we judge others' knowledge and competence. These assessments are neither objective nor consistent.[1]

As described in Chapter 3, perceiving a social interaction involves categorizing individuals into a salient social group, which is associated with various features, stereotypes, and social biases. These associations influence how we decide who is an epistemic peer, inferior, or superior before we even evaluate their evidence base or reasoning abilities. Simply in virtue of being part of a particular social category we may upgrade or downgrade a person's knowledge or competence in a certain domain. For example, psychologists have discovered a set of universal dimensions along which people perceive individuals: warmth and competence (Fiske, Cuddy, and Glick 2007). That is, we perceive individuals as warm or cold and competent or incompetent.[2] Across many different cultures, including collectivist and individualist societies, adults of all ages tend to spontaneously and implicitly associate elderly women with warmth and incompetence (Cuddy, Norton, and Fiske 2005; Fiske, Cuddy, and Glick 2007; Hummert et al. 1994). This is evidence that many people tend to habitually downgrade the epistemic status of elderly women simply in virtue of their social category. Of course, we can override implicit associations, but in many cases, this will require that the evidence be obvious that the elderly woman is not ignorant or incompetent, e.g., she may be a famous scholar on some topic. In the absence of such clear and overt evidence, to see an elderly woman as a contender to be a peer or superior, we must override our *default assumption* about the warmth and incompetence of elderly women. Doing this seems to require awareness of the associations and their effect on one's behavior, attention, and cognitive effort. For this reason, such associations are difficult to excise from one's judgments.

In-group/out-group status also significantly affects our judgments of other people's epistemic status. As discussed in Chapter 3, we usually have more favorable attitudes toward and empathize more with in-group members, especially people who share our gender, race, age, religion, or nationality, than toward people do not share these features. The phenomenon of in-group favoritism suggests that we are generally less likely to regard out-group members as epistemic peers or superiors. There are at least two ways in which such patterns may be upended. First, there may be clear and undeniable evidence that an out-group member is an epistemic peer or superior with respect to some domain. Second, we may regard an out-group as generally dominant in a certain domain, and our in-group as subordinate in that domain. For example, we may have positive stereotypes about an out-group's knowledge and competence in math, e.g., the stereotype Asians are good at math. Or we may have negative stereotypes about our in-group's knowledge and competence in math, e.g., the stereotype that women are not good at math. However, setting aside specific dominant/subordinate stereotypes, in general we tend to upgrade the status of our in-group and either downgrade the status of out-groups or give deflationary accounts of their seeming competence (e.g., they are not that smart; they just work hard). We are especially subject to these patterns of bias in the context of threat or competition, like when people disagree about some important issue or scarce resources are at stake.

Self-enhancing biases and other-downgrading biases make for a dismaying assessment of how we judge others' knowledge and competence. The evidence reviewed indicates that we are most likely to regard another person as an epistemic peer when she is part of our relevant in-group and in fact she is moderately epistemically superior. In most other conditions, other things being equal, we are likely to regard epistemic peers as inferior, and we are likely to regard moderately epistemically superior out-group members as peers or even inferior. These assessments may be modified with further deliberation. First impressions have a way of sticking around and influencing even careful deliberation, however (Willis and Todorov 2006).

Mindreading judgments are crucial to our assessments of others' knowledge and competence. When we judge that someone is an authority on, say, climate change, we judge that she knows what she is talking about when she says that global temperatures are trending upward – that is, she has deep background knowledge upon which she is drawing – and her apparent confidence in her assessment is based on possessing such knowledge. As this schematic example illustrates, judgments about personality traits, mental states, and epistemic states are interconnected. Default assumptions about the knowledgeability and competence of certain kinds of people with respect to various topics will influence the kinds of mental state and personality trait inferences we make about individuals. And these mental state and personality trait inferences will tend to confirm the default assumptions about knowledgeability and competence. Put a different way, we construct and employ holistic models of others that help us make sense of how people think and act. When our starting assumptions lead us to adopt skewed models, unless there is glaring evidence of a mismatch between the person's behavior and the model, we will tend to interpret the rest of the behavior to be in line with the model and thereby seemingly confirm the appropriateness of the model for that person. We tend to employ skewed models of ourselves and others in the conditions described above. Because of stereotypes and various associations with social categories, our models are especially skewed toward underestimating out-group members, and this tendency is strongest when we are ignorant or incompetent in the relevant topic.[3]

The processes I have described influence our default assumptions about who is knowledgeable and competent about various topics, and they also influence how we reason about evidence of our own and others' competence and knowledge. Confirmation bias, our tendency to look only for evidence that confirms our ideas, further entrenches these patterns. The somewhat depressing conclusion is that we often are pretty poor judges of others' knowledge and competence when the evidence is not abundantly obvious. This is problematic for the epistemology of peer disagreement discussions, which often assume that we are reliable judges of who is an epistemic peer. This is not just a theoretical problem, however. There are real practical costs to our tendency to be poor judges of others' knowledge and competence. Individuals who are comparatively less knowledgeable and competent may be hired, elected, or promoted over individuals who would do a superior job. There are costs to the knowledgeable and competent individuals who are rejected, of course. But there also are costs to the employers, constituencies, and stakeholders in general when relatively incompetent, ignorant individuals gain power and influence. There is no simple formula to override these default assumptions and skewed reasoning, though the first requirement is awareness of their effects on our judgments. These theoretical and practical concerns ought to push us more toward intellectual humility (Hazlett 2012, 220; Roberts and West 2015).

6.3 Application to the epistemology of peer disagreement

I have cast doubt on our ability to accurately judge others as epistemic inferiors, peers, and superiors. What follows is a consideration of the debate about what to do when someone you take to be an epistemic peer disagrees with you. Disagreement between epistemic peers has received considerable attention recently in the epistemology literature. The puzzle of peer disagreement arises when someone I acknowledge as an epistemic peer disagrees with me. When I regard you as my epistemic peer with respect to X but we come to different judgments about X, what is the rational thing for me to do? Perhaps it is rational for me to remain steadfast in my judgment about X (Kelly 2011). Or perhaps the rational thing to do is conciliate and reduce my confidence in my judgment (Christensen 2007). Proponents of the steadfast view and the conciliation view hold that such disagreements amongst acknowledged epistemic peers are widespread. Thus, proponents of both views argue that the debate has broad implications for the rationality of retaining many of our controversial beliefs.

The assumption that this kind of peer disagreement is widespread is dubious. The conception of an epistemic peer is an idealization that rarely obtains in the real world. It is rare that two interlocutors are, or will regard each other as, equally intelligent, equally free from bias, equally competent reasoners, and equally informed about some issue, especially when that issue is even somewhat complex. Nathan King (2012) rigorously and persuasively argues that most real-world disagreements are not disagreements between acknowledged epistemic peers. Philosophers in the peer disagreement debates recognize that their examples are idealized and simplified examples of peer disagreement, but these examples are supposed to make our intuitions clear. Philosophers in this debate argue that the lessons we learn from the simplified cases apply to real-life complex cases as well if we use the looser standard of *comparable* rather than *equal* evidence, reasoning ability, etc. (Christensen 2007, 211–212; Kelly 2011). The lessons from the previous section seem to indicate that even judgments of epistemic parity are likely to be rare, though certainly not as rare as judgments of epistemic equality. In any case, from here on out, I will understand an epistemic peer to be someone who is epistemically comparable. I shall focus on the discussion of what to do when someone we judge to be *comparable* in terms of evidence and knowledge, perceiving and reasoning ability, and freedom from bias disagrees with us.[4]

With that caveat in mind, what should we do when we encounter a case of peer disagreement? The steadfast view (Kelly 2011) and conciliatory view (Christensen 2007), as they are typically debated, hinge on whether it is legitimate to use one's own reasoning about a topic in evaluating a peer's reasoning about that topic (Christensen 2011). That is, when an epistemic peer disagrees with you, can you use the fact that she disagrees with you as evidence that she is less likely to be correct? The steadfast view says yes, you can, and that tips the scales in favor of your own judgment. The conciliatory view says that you should not, and without that resource we must conclude that your epistemic peer has just as much reason to think she is correct. In that case, it is rational to reduce confidence in your judgment. That is an interesting debate worth having, but it seems to me that the psychology of disagreement gives us a pretty straightforward answer about what to do when someone you take to be an epistemic peer with respect to some topic disagrees with you.

Given what we know about our tendency for self-enhancing biases and other-downgrading biases, especially for out-group members who disagree with us, it seems that the rational thing to do is to conciliate and reduce confidence in our own judgments rather than remain steadfast. After all, the psychological data suggests that if we have judged someone to be epistemically comparable to us about X (especially if they disagree with us about X), they are likely to be at least and very possibly *more* knowledgeable and competent with respect to X. Indeed, if we disagree with someone we take to be an epistemic peer who is a member of a subordinate out-group, she is likely to be *at least* moderately epistemically superior.[5] In both of these cases, it seems like the safest, most reasonable response to what we take to be peer disagreement is reducing confidence in our own judgments. We should even take pause when we disagree with someone we take to be a moderately inferior epistemic agent from a subordinate out-group, because it is likely that we have underestimated her knowledge and skills. Thus, heeding the advice above for more intellectual humility pushes us toward conciliation in many cases of disagreement (Hazlett 2012).

Let's scale out a bit and consider the epistemology of peer disagreement debates more generally. The allure of the peer disagreement discussion is that it promises to tell us what we rationally ought to do in the face of disagreement with another intelligent, informed, well-meaning individual. The literature is a bit hamstrung by the fact that the idealizations about epistemic peers are problematic and this limits its applicability to real-life disagreements. But even if we find a way to resolve that issue, it seems that the glaring problem is the lack of focus on *how* we decide who is an epistemic peer. The debate over peer disagreement glosses over this thorny issue.[6] This is a curious fact, because deciding who is an epistemic peer is central to the epistemology of disagreement. Nevertheless, this is exactly where mindreading theorists can step in and help. If the view I argue for in this book is correct, then there is not a hard and fast distinction between folk psychology and folk epistemology. These domains actually are closely connected in the sense that mindreading processes are deeply involved in assessing the knowledge and competence of others.

A comprehensive theory of mindreading that takes into consideration all the various ways in which we understand and interact with others can shed light on how we evaluate others' knowledge and competence. My analysis of the epistemology of peer disagreement is, of course, subject to debate. If, like the steadfast theorists, one holds that you can rely on your own reasoning about X to evaluate a peer's reasoning about X, and one assigns significant weight to one's own reasoning, this may tip the scales in a different direction even with the recognition that our evaluations of others' knowledge and competence often are biased. The take-home lesson is more about the applicability of mindreading to topics in social epistemology than about coming to a particular conclusion about whether to remain steadfast or conciliate in the face peer disagreement. Though I strongly believe we would all do well to adopt a bit more intellectual humility, the conclusion I most want epistemologists and mindreading theorists to embrace is that mindreading is highly relevant to debates in social epistemology.

6.4 Application to epistemic justice

Epistemic injustice is a topic at the intersection of ethics, epistemology, and feminist philosophy. Epistemic injustice harms potential knowers by denying them epistemic resources or credibility. Testimonial injustice is a form of epistemic injustice that is particularly relevant here. Testimonial injustice occurs when a hearer's prejudices discount a speaker's credibility. In the rest of this chapter, references to epistemic injustice should be understood to refer to testimonial injustice specifically.

My discussion of how we judge others' knowledge and competence is especially relevant to the discussion of epistemic injustice. Unlike the literature on the epistemology of peer disagreement, the literature on epistemic injustice does not overlook the empirical and theoretical work on how social dynamics mediate our interactions. In fact, this work is a central focus of debates about epistemic injustice. For example, epistemologists, particularly those in social epistemology, focus on how epistemic agents' social worlds influence their status as knowers. This work is deeply influenced by some of the empirical work described in this book. Virtue epistemologists also work on epistemic injustice, and their focus tends to be on how to cultivate the virtue of epistemic justice and avoid the vice of epistemic injustice. Feminist philosophers who work on epistemic injustice theorize about the gendered and racial patterns of epistemic injustice and ways to intervene or mitigate the effects of epistemic injustice. Clearly the philosophers working on this topic are aware of much of the relevant empirical evidence on how extant social dynamics mediate our interactions with others. Thus, my discussion here will focus less on detailing the relevance of these data and more on interventions likely to succeed in light of the view I have developed in this book.

Model Theory plausibly explains when and how epistemic injustice is likely to arise, and we can use the theory to evaluate how successful various interventions are likely to be. As discussed previously, Model Theory can explain the tendency to upgrade the epistemic status of individuals from certain social groups and downgrade the epistemic status of individuals from other social groups. On the view I develop, epistemic upgrading and downgrading is not, or need not be, a deliberate choice. Rather, given our background stereotypes and social biases and the subtle and complex interaction between situational context, personality trait inferences, and mental state inferences, we tend to just *see* some people as more authoritative and competent than others. Indeed, it is likely that when we are challenged on our perceptions of authority, competence, and lack thereof, we will sincerely reference the features we have noticed that justify these perceptions and double down on our epistemic judgments. Put in Model Theory terms, social categorization patterns and our goals determine the character of the models we construct and how we employ them, and once we employ a model it is cognitively efficient to interpret further evidence in light of that model. Employing a different model will happen only if there are glaring inconsistencies or if we care deeply about getting it right and suspect that we have employed an inappropriate model.

This analysis of how epistemic injustice arises has implications for how successful various interventions on epistemic injustice will be. Fricker (2003, 2007) argues that to combat epistemic injustice individuals should train themselves to be vigilant about their epistemic prejudices and compensate for these prejudices by according more epistemic authority to individuals from traditionally subordinate groups. Similarly, Jose Medina (2013) argues that we should fight epistemic injustice by developing an epistemic sensibility that detects and corrects both undeserved credibility deficits and undeserved credibility excess.

In light of the view developed in this book, I think both of these interventions are unlikely to succeed. Both interventions rely on individuals to police themselves, i.e., to get themselves to notice when they are being biased and compensate for those biases. In the vast majority of cases of epistemic injustice, people do not see themselves as biased. Their reasoning seems appropriate because all of the information they are attending to confirms the legitimacy of their reasoning. The problem is that they are not attending to the relevant information that would highlight the fact that their reasoning is biased. Fricker and Medina's advice may work in some very obvious cases of biased reasoning where it takes little effort to notice disconfirming information. This is the low-hanging fruit. But even in those cases, one would tend to notice one-off mistakes, not a pattern of biased reasoning. Recognizing a pattern of biased reasoning would be much harder than recognizing obvious one-off mistakes because of our tendency to attend to information that confirms our perspectives and our hesitancy to think of ourselves as biased. Thus, Fricker and Medina's interventions likely will fail to effectively combat epistemic injustice. Indeed, these suggestions may even backfire. We have powerful tendencies to think of ourselves as unbiased and simply perceiving the world how it is (i.e., Naïve Realism) and to consider information that confirms that. Given these tendencies, we may take ourselves to follow Fricker and Medina's advice and conclude happily that we have done our due diligence in fighting epistemic injustice when, all the while, we have not even recognized our biases. Such reasoning patterns would likely make us more confident in our own judgments. Following this advice thus may be counterproductive for the goal of inhibiting epistemic injustice.[7]

So, what *is* likely to work? Answering this question requires understanding the impediments to recognizing our own epistemic injustices. These impediments include stereotypes and implicit associations we do not recognize we have or do not see as inappropriate, false, or harmful; our tendency to regard those like us (*especially* if we are part of a dominant social group) as more authoritative and competent; our tendency to regard ourselves as unbiased; and confirmation bias.

Even though confirmation bias makes epistemic injustice very difficult to counter, it is unhelpful to target interventions toward confirmation bias. Confirmation bias is incredibly hard to resist. We are subject to confirmation bias even when we know the source of the information is unreliable and even when accuracy is incentivized. We are especially subject to confirmation bias when our views are challenged.[8] Facts do matter of course, and usually we do not simply disregard well-established facts (Wood and Porter 2018). The trouble with social interactions is that we are dealing with interpretations rather than well-established facts, so confirmation bias is very sticky in this domain. For this reason, it is better to aim interventions at the other impediments to recognizing our own epistemic injustices.

What will help us fight stereotypes, implicit associations, in-group favoritism, and Naïve Realism from manifesting in epistemic injustice, along with a host of other ethical transgressions? There is no silver bullet, which may be disappointing but not surprising to those who have read this far into book. However, there are several behavioral interventions that are promising for mitigating the effects of these social biases.

The first kind of intervention is to educate oneself on commonly held stereotypes and implicit associations that are inappropriate, false, or harmful. It is not always easy to know one's own stereotypes or implicit associations because these are context-sensitive associations that one employs on the fly, so to speak. We call on them when we need them, and they usually are not subject to careful introspective scrutiny. For this reason, I suggest learning about the stereotypes that others have, especially people in one's professional and demographic circles. This requires reading and listening to scholars who work on these topics, people who have first-hand experience of bias and microaggressions underwritten by stereotypes and implicit associations, and stepping outside one's typically homogeneous social circles to have *meaningful* interactions with people with different social experiences. By meaningful, I mean that you engage in conversations about your different social experiences and learn how the world looks from a different perspective.[9] In addition, one must understand what situations are likely to evoke stereotypes and other social biases and form intentions for how to act to in those situations and practice acting that way.

The interventions described above are partly about educating oneself about social biases. Such education is not sufficient on its own to stop one from employing these stereotypes and associations on the fly, of course. But the first step to stopping these patterns of inference is understanding them. We need to know that most people like us have a default assumption that, for example, elderly adults are less competent than non-elderly adults. The other aspect of these suggestions concerns changing one's social environment, e.g., the perspectives one regularly hears, the news sources one is exposed to, the individuals one comes to think of when thinking about various social groups, etc. We all rely on our social environments to shape our social interactions. We pick up enormous amounts of socially relevant information from our social environments without explicit instruction. Most of the associations and stereotypes we learn we implicitly detect from our social environments. Based on how individuals around us and our subculture in general treat various situational contexts, certain kinds of social interpretations become our default for those contexts. None of this requires deliberate teaching. We simply detect these patterns of inferences from our social environments and adopt them with little critical reflection. Thus, the character of one's social environment significantly affects the kind of inferences one is likely to make. This is why it is important both to learn about ethically problematic patterns of inference *and* work to change one's social environment to better support one's quest to think and behave justly (Huebner 2016).

To construct effective interventions, we also need to understand *when* and *how* we are likely to commit epistemic injustices and other ethically problematic patterns of reasoning. When efficiency is a primary motivation in our social interactions, we tend to attribute stereotypes to out-group members and project our own perspectives to in-group members. The ethically problematic patterns of inference are likely to go undetected when we are using such heuristics to quickly interpret social interactions. We cannot avoid using such heuristics; careful deliberation is not an option in every situation, nor is careful deliberation a guarantee of objective reasoning. However, in contexts where we know that we are likely to employ such heuristics – when we are under heavy cognitive load already, when we are busy, or when the situation seems very familiar to us – we should exploit structural and institutional measures to serve as a check on potential biases.

Consider, for example, hiring and academic grading. In both cases, individuals making decisions often are highly motivated to be efficient in their cognitive efforts. Using software to anonymize job résumés and academic assignments is one way to combat the tendency to rely on problematic stereotypes and in-group projections. When résumés and tests are not anonymized, it is easy to think charitably about the work of candidates and students who have similar backgrounds and not give candidates and students with different backgrounds a similar benefit of the doubt. Indeed, in cases where we are speeding through piles of résumés and tests, social stereotypes about who is likeable, smart, competent, and working are bound to influence our quick judgments. Anonymizing is one way to counteract that influence. In the hiring process, once candidates get to the interview stage, having pre-structured interviews where all the questions are written down in advance and known by all the interviewers, and where each interviewee gets all and only the same questions is another way to avoid subtle influences of stereotypes and projections in hiring decisions. For grading, consistently applying structured and specific grading rubrics to anonymized assignments can help counteract the influence of problematic stereotyping and projecting.

Epistemic injustice and other ethically problematic inferences also arise for deliberative reasoning, of course. With education, social change, and open, frank conversation about the potential for bias, these effects can be lessened. For example, in the later stages of deciding whom to hire, individuals on job search committees probably are employing deliberative reasoning. They are debating the merits of each candidate, trying to figure out who would be the best fit, who is likely to stay in the job for the long term, etc. Bias surely can seep in here, so it is important to openly discuss the possibility of bias with others (who hopefully also care to avoid bias). Doing this allows others in the group to flag potentially problematic reasoning about candidates when you do not notice the problematic reasoning. In general, exploiting one's social circles to help one recognize and intervene on ethically problematic patterns of reasoning is key to avoiding social biases. Without such externally supported reasoning, all of the above advice is likely to fail because one can so easily fall back into unchecked, ethically problematic thinking.

The interventions suggested here differ from Fricker and Medina's interventions in that they do not conceive of epistemic justice as a stable trait that one can develop and employ in any situation. On the view I develop, being epistemically just requires different strategies in different situations, and being committed to epistemic justice in one context does not entail that such a commitment will be important or effective in other contexts. My suggestions rely heavily on altering and exploiting one's social environment to police epistemic injustice.[10] Fricker and Medina's suggestions, in contrast, primarily involve self-policing, which I have argued is likely to fail or even backfire given what we know about our tendency toward motivated reasoning.

We have good prima facie empirical reasons to think that these interventions, applied consistently and in good faith, are likely to help mitigate the effects of epistemic injustice and other ethically problematic forms of reasoning. They will not eliminate all social ills, though. There is no single intervention or even set of interventions that can do that. The best we can do is to continue to have conversations about what in fact contributes to these social ills and how to effectively combat them given what we know about how we think, feel, and behave in various contexts.[11]

6.5 Assessment

This chapter has focused on the relevance of the broader conception of mindreading to debates about peer disagreement and epistemic injustice interventions. In this assessment section, I will take a step back and consider what the views developed in this book suggest about mediating controversial topics more generally. The evidence canvassed in this book indicates that in contentious debates, we are likely to be unreceptive to ideas we disagree with when they are advanced by a speaker who is saliently different from us. In such cases, we are likely to discount the credibility of the speaker and/or simply dismiss the idea. This is both totally unsurprising and completely discouraging. How will we ever make dialogical, let alone practical, progress on contentious issues if we have the default tendency to dismiss ideas we disagree with from speakers who differ from us?

I have no idea how to construct effective discussions with very large groups of people, so I have no expert advice to offer those hoping to engage in national or international dialogues. However, the perspective on mindreading adopted in this book does have some insight for mediating debates at a more personal level. The key seems to be establishing relevant commonality with each other.

Consider as an example how effective pedagogy works. When we teach our students, we are in many ways very different from them. They are from a different generation, they often have different backgrounds, interests, values, and beliefs, and they occupy a very different institutional role in this context than we do. As philosophy teachers, we expose students to ideas and arguments that challenge their values and beliefs. Given some of the data canvassed in this book, one would expect that, in response, most students would discount our credibility or simply dismiss the ideas we present to them. Often it does not turn out that way, though. Students more or less thoughtfully engage with controversial ideas we present. This engagement likely is *not* due to the charming nature of philosophy instructors, as we are not generally known as a charming lot. I think the explanation has to do with establishing a shared goal of understanding and evaluating the controversial ideas. Effective teachers of controversial material do not simply tell students what to believe or profess their own beliefs and call it a day. Rather, they jointly engage in the project of assessing these controversial ideas. In engaging with students in this way, we mark ourselves as part of the same in-group as the students.

There are always differences between the students and the professor, just as there are always differences amongst the students themselves, and neither the students nor the teachers lose track of those differences. However, in-grouping and out-grouping is flexible and sensitive to context. Who counts as part of a relevant in-group depends on perceived similarity of goals, experiences, or beliefs and values. The ideas discussed in this book suggest that we respond quite differently to people we perceive to be part of a relevant in-group from those we perceive to be part of a relevant out-group. We like in-group members more than out-group members, and we interpret them more charitably than out-group members. This is the case for age, race, and gender, but also more transient groupings like teams, projects, and even arbitrarily delineated groups.

In a pedagogical context, effectively teaching controversial material often involves establishing shared goals and thereby establishing ourselves as part of an in-group with the students. This makes the discussion of controversial ideas less threatening because it is not coming from someone with radically different goals with respect to this material. (For example, if students suspected that you were simply trying to convert them to a different belief, thoughtful, open engagement would be much more difficult to achieve.) This all makes sense in light of the data on the effects of in-grouping and out-grouping.

The importance of establishing an in-group with students in teaching applies to effectively mediating contentious disputes outside of the classroom, as well. When we are trying to engage in a debate with someone who strongly disagrees with us about some issue, it is important to mutually establish some relevant commonality, e.g., a common goal, shared values or beliefs, or similar experiences. This commonality is lubricant for the discussion; it reduces friction and helps both parties sincerely engage in the conversation. It does *not* ensure eventual agreement; nothing does that. But it does allow for more thoughtful engagement about difficult issues. If both parties sincerely regard each other as having relevant commonalities, it is easier for both parties to consider how someone, who is in some ways quite similar to them, could come to a really different conclusion about the matter at hand.

It would be out of keeping with the spirit of this book to end on a Pollyannaish note. Mediating contentious discussions is difficult even when all the parties see each other as having the similar goals and values. Often what matters is how you rank your various goals and values in particular contexts. NRA members and proponents of gun control both value liberty and are motivated by personal and societal safety concerns. Recognizing these shared values and goals will not dissolve their disagreements. But if their recognition is sincere, it is useful. At the very least it will allow them to move past the obvious *ad hominem* and straw man arguments. And that would be progress.

Notes

1 In other work, I address evidence of what I call *mindmisreading*, i.e., inaccurate mindreading judgments. Here I will focus on the inconsistency and subjectivity of these judgments, not their accuracy. See Spaulding (2016) for more on mindmisreading.

2 Our competence judgments seem to be based on the target's social status, and our warmth judgments seem to be based on whether the target represents competition or threat (Cuddy, Norton, and Fiske 2005). The dimensions of warmth and competence combine to generate distinctive emotions (Fiske, Cuddy, and Glick 2007). Stereotypes that involve warmth and incompetence tend to evoke pity in observers. Stereotypes that involve warmth and competence tend to evoke admiration in observers. Stereotypes that involve coldness and competence tend to evoke envy in observers. Stereotypes that involve coldness and incompetence tend to evoke contempt in observers.

3 This analysis generates the depressing prediction that the climate change expert will have little success convincing an ignorant climate change skeptic.

4 Epistemic comparability is not without its own problems. Weakening the knowledge and competence requirements from equal to comparable introduces a fair bit of context sensitivity. Context sensitivity makes it difficult to construct useful principles for deciding when we have a case of epistemic comparability. Without guiding principles, almost everyone could count as epistemic peers. But if that's the case, disagreements amongst epistemic peers will be extremely heterogeneous, which threatens the idea that we can give definitive yet general guidance on what is rational to do in a case of peer disagreement. Going the other direction, making the notion of epistemic peers quite conservative, limits the real-world applicability of these debates. So, theorists in this field will have to work out some sort of decision criteria for who counts as epistemically comparable.

5 This, of course, is relative to the domain and the dimensions of in-grouping/out-grouping. Surely the fact that someone is, say, Buddhist and thus not part of my religious in-group has no bearing on my judgments about whether he is likely to be correct about his answer to a math problem. When I discuss in-grouping and out-grouping with respect to peer disagreement, the relevant dimensions of in-grouping and out-grouping are ones that are associated with the domain of disagreement. This may be a very close association, as it is when we are in a disagreement about guns with someone of a different political affiliation, or it may be a subtler association, as it is when we disagree with someone of a different gender about the answer to a math problem. Thus, the idea is that when we disagree with someone we regard as part of some *relevant* out-group, we should be wary of our assessments.

6 Nathan Ballantyne (2015), a recent exception to this trend, considers the biases we attribute to others in order to debunk their testimony. Ballantyne argues that our methods for attributing bias are themselves makeshift and biased, which

indicates that debunking dissenting testimony is not a reliable method for responding to peer disagreement. He argues that this is evidence in favor of the conciliation view.

7 See Sherman (2016) for an argument that we should not think of epistemic justice as a virtue. Sherman's arguments are different from mine in their theoretical commitments and goals, but there is some overlap in our analyses of why Fricker's interventions are likely to fail and potentially even backfire.

8 For more on confirmation bias, see Chapter 4.

9 Typically, people concerned with combatting their own biases are comfortable with the first two suggestions, but I find that almost everyone finds the last suggestion challenging for lots of reasons, e.g., awkwardness, perceived lack of opportunity, concern about tokenism and insincerity, time investment, etc. It is important to understanding the lived experiences of people from different social backgrounds though, and it is important that we conceptualize social groups as made up of individuals with differences, similarities, idiosyncrasies, hopes, ambitions, fears, etc. We are very good at thinking about our own in-groups this way. This nuanced view of out-groups may be achieved by reading and listening to different social groups from afar, but it does help to get to know people from different social groups.

10 Holroyd and Sweetman (2016) offer a useful distinction between insulating vs. mitigating strategies for combatting bias. Alfano (2015), Davidson and Kelly (2015), and Washington (2016) also offer accounts that exploit the social environment to help check epistemic injustice.

11 As usual, these discussions are with and about individuals who already care about fighting biases. Such discussions leave out individuals who do not care to know about or fight such biases. As a result, these interventions are bound to have limited effects overall. Different contexts of bias require different interventions, so there is no such thing as a general, let alone simple, solution to biased thinking and behavior.

References

Alfano, M. 2015. "Becoming less unreasonable: A reply to Sherman." *Social Epistemology Review and Reply Collective* 4 (7): 59–62.

Ballantyne, N. 2015. "Debunking biased thinkers (including ourselves)." *Journal of the American Philosophical Association* 1 (1): 141–162.

Christensen, D. 2007. "Epistemology of disagreement: The good news." *The Philosophical Review*: 187–217.

Christensen, D. 2011. "Disagreement, question-begging, and epistemic self-criticism." *Philosophers' Imprint* 11 (6): 1–22.

Cuddy, A. J. C., M. I. Norton, and S. T. Fiske. 2005. "This old stereotype: The pervasiveness and persistence of the elderly stereotype." *Journal of Social Issues* 61 (2): 267–285.

Davidson, L., and D. Kelly. 2015. "Intuition, judgment, and the space between: A reply to Sherman." *Social Epistemology Review and Reply Collective* 4 (11): 15–20.

Fiske, S. T., A. J. C. Cuddy, and P. Glick. 2007. "Universal dimensions of social cognition: Warmth and competence." *Trends in Cognitive Sciences* 11 (2): 77–83.

Fricker, M. 2003. "Epistemic justice and a role for virtue in the politics of knowing." *Metaphilosophy* 34 (1–2): 154–173.

Fricker, M. 2007. *Epistemic Injustice: Power and the Ethics of Knowing*. Oxford: Oxford University Press.

Hazlett, A. 2012. "Higher-order epistemic attitudes and intellectual humility." *Episteme* 9 (3): 205–223.

Holroyd, J., and J. Sweetman. 2016. "The heterogeneity of implicit bias." In *Implicit Bias and Philosophy*, edited by Michael Brownstein and Jennifer Saul. New York, NY: Oxford University Press.

Huebner, B. 2016. "Implicit bias, reinforcement learning, and scaffolded moral cognition." In *Implicit Bias and Philosophy*, edited by Jennifer Saul and Michael Brownstein. New York, NY: Oxford University Press.

Hummert, M. L., T. A. Garstka, J. L. Shaner, and S. Strahm. 1994. "Stereotypes of the elderly held by young, middle-aged, and elderly adults." *Journal of Gerontology* 49 (5): P240–P249.

Kelly, T. 2011. "Peer disagreement and higher order evidence." In *Social Epistemology: Essential Readings*, edited by Alvin Goldman and Dennis Whitcomb, 183–217. Oxford: Oxford University Press.

Kennedy, K. A., and E. Pronin. 2008. "When disagreement gets ugly: Perceptions of bias and the escalation of conflict." *Personality and Social Psychology Bulletin* 34 (6): 833–848.

King, N. L. 2012. "Disagreement: What's the problem? Or a good peer is hard to find." *Philosophy and Phenomenological Research* 85 (2): 249–272.

Kruger, J., and D. Dunning. 1999. "Unskilled and unaware of it: How difficulties in recognizing one's own incompetence lead to inflated self-assessments." *Journal of Personality and Social Psychology* 77 (6): 1121.

Medina, J. 2013. *The Epistemology of Resistance: Gender and Racial Oppression, Epistemic Injustice, and the Social Imagination*. Oxford: Oxford University Press.

Miller, D. T., and M. Ross. 1975. "Self-serving biases in the attribution of causality: Fact or fiction?" *Psychological Bulletin* 82 (2): 213.

Pronin, E. 2007. "Perception and misperception of bias in human judgment." *Trends in Cognitive Sciences* 11 (1): 37–43.

Pronin, E., D. Y. Lin, and L. Ross. 2002. "The bias blind spot: Perceptions of bias in self versus others." *Personality and Social Psychology Bulletin* 28 (3): 369–381.

Roberts, R. C., and R. West. 2015. "Natural epistemic defects and corrective virtues." *Synthese* 192 (8): 2557–2576. doi:10.1007/s11229-015-0669-5.

Sherman, B. R. 2016. "There's no (testimonial) justice: Why pursuit of a virtue is not the solution to epistemic injustice." *Social Epistemology* 30 (3): 229–250. doi :10.1080/02691728.2015.1031852.

Spaulding, S. 2016. "Mind misreading." *Philosohical Issues* 26 (1): 422–440.

Washington, N. 2016. "I don't want to change your mind: A reply to Sherman." *Social Epistemology Review and Reply Collective* 3: 10–14.

Willis, J., and A. Todorov. 2006. "First impressions: Making up your mind after a 100-ms exposure to a face." *Psychological Science* 17 (7): 592–598. doi:10.1111/j.1467-9280.2006.01750.x.

Wood, T., and E. Porter. 2018. "The elusive backfire effect: Mass attitudes' steadfast factual adherence." *Political Behavior*. doi:10.1007/s11109-018-9443-y.

7 Concluding thoughts

In this book, I offer a partial defense of the mainstream view of mindreading. In Chapter 2, I argue that we regularly attribute mental states in order to understand, anticipate, and manipulate others' behavior. However, I also argue that mindreading encompasses a much broader set of phenomena than is acknowledged by mainstream mindreading theories. Chapters 3 and 4 canvass the complexity and diversity of the inputs to mindreading, our approaches to mindreading, and the kind of judgments we end up producing. A complete and adequate account of mindreading ought to explain the various modes of mindreading, why we use them, and how they break down. In Chapter 5, I offer Model Theory as a particularly promising comprehensive account of mindreading. Model Theory is broad and flexible enough to explain the diversity of modes of mindreading, and yet it also makes specific empirically testable predictions. The view of mindreading I lay out is useful beyond the narrow confines of the mindreading debate. In Chapter 6, I apply the work on mindreading to debates in epistemology and ethics. I argue that the view of mindreading I develop has important implications for peer disagreement and interventions on epistemic injustice.

In some ways, the book presents a fairly cynical analysis of our social interactions. After all, it focuses considerably on the effects of stereotypes, social biases, and self-motivated reasoning. I argue that these often lead to mindreading errors, circular reasoning, and biased thinking and behavior. This perspective does not exactly sing the praises of our mindreading abilities. I see this, in part, as a corrective to an overly simplistic and optimistic view of our mindreading practices. The standard story about mindreading vastly underestimates the complexity, diversity, and messiness of mindreading. Serving as a corrective is not the only reason for the pessimism, though. In reality, there are numerous ways in which subtle and not-so-subtle mind-misreading undermines social interaction. I discuss examples of this in Chapter 5.

Our social cognitive capacities evolved to be useful in small groups of similar and often genetically-related people. In that evolutionary context, it makes sense to treat people in your small, homogenous social group very differently from people outside that group. Outsiders almost always presented competition and threat. Thus, it is easy to see how in-group/out-group dynamics could have been adaptive in that evolutionary context. However, we no longer live in very small, homogenous groups of people. Quite recently in evolutionary history, we have shifted from very small, homogeneous groups to large, geographically broad, and demographically diverse social circles. We now regularly encounter people of different races, nationalities, religions, different cultural practices, expressive behaviors, and social norms. We interact with an incredibly diverse array of people. From our neighbors, coworkers, social media contacts, to celebrities we follow, musicians we love, and actors we watch, we regularly encounter people who are demographically different from us. This dramatic social expansion is wonderful in many ways of course, but it brings with it many challenges. Meeting these challenges requires an honest assessment of our cognitive tendencies. That is another reason why it is important to point out our limitations and explain our propensities for biased social reasoning. Thus, I hope both theorists interested descriptive accounts of our social practices and theorists interested in normative analyses of social injustice heed the lessons learned here.

Looking forward, I see many exciting areas of research for mindreading theorists to investigate. Perhaps most obviously, there is more work to be done on constructing and developing properly comprehensive general accounts of mindreading. I offer Model Theory as one promising general and comprehensive account, but really the account is relatively schematic at this point. Thus, one avenue for future research is developing, testing, and refining Model Theory, expanding existing theories to accommodate the phenomena canvassed here, or coming up with alternative general theories of mindreading.

In addition to constructing general theories of mindreading, the ontogenetic development of this broader conception of mindreading promises to be particularly interesting. Research on the development of mindreading typically proceeds independently of the research on the development of social categorization and social preference. Recently, however, empirical and philosophical work has begun to connect these two social developments in the study of social cognition (Abrams and Rutland 2008; Westra 2017).

Gaze cueing, i.e., following the gazes of other people, is an important mechanism of joint attention, which is foundational for mindreading. Recent empirical evidence suggests that infants as young as 7 months old preferentially look at the gazes of own-race adults to guide learning and social interactions (Xiao et al. 2017). Within the first year, infants begin to form preferences for own-language speakers and expect that people who speak the same language will cooperate, have similar customs, and like the same foods (Liberman, Woodward, and Kinzler 2017). After the first year, infants selectively imitate in-group members (Buttelmann et al. 2013). By 5 years old, children perceive less humanness in faces of out-group members (McLoughlin, Tipper, and Over 2017) and show in-group favoritism and out-group bias even for arbitrary, minimal groups (Dunham, Baron, and Carey 2011). It could be a coincidence that folk psychological development from infancy to preschool years goes through massive changes at the same time that social categorization and social preferences develop, but there is good theoretical reason to think these social practices develop in tandem (Abrams and Rutland 2008). Indeed, there is some evidence that the development of in-group favoritism and out-group bias is strongly related to developing social cognitive skills (Aboud 2003; Abrams et al. 2014).

Thus, in the future, it will be fruitful and interesting to look at the developmental trajectories and interactions between mindreading and social biases. For example, it would be interesting to learn how children conceive of different social groups (as natural kinds or merely conventional groups), how they learn and reason about these groups, the kind of mindreading inferences they are likely to make about members of social groups, and how culturally malleable these developments and processes are.[1]

A related future area of research concerns the distinctive social inferences of people who have Autism Spectrum Disorder. If, as I speculate above, social categorization and social preference develop in tandem, we should expect to see distinctive patterns of social categorization and social preference for people on the Autism Spectrum. Investigating how such individuals socially categorize, form in-groups and out-groups, and base behavioral inferences off in-group/out-group judgments would significantly enhance the understanding social cognition for individuals with Autism Spectrum Disorder. There is currently very little work on these issues, so it is an area ripe for empirical and theoretical study.

The nature of the explanations we produce in various modes of mindreading is another promising area of research. Although there is a rich body of empirical and philosophical work on the nature of explanation and understanding, much of this work overlooks or does not quite fit for mindreading. Mindreading explanations tend to be both causal and normatively laden (e.g., rationalizing, justifying, or blaming). They can be teleological, formal, or even mechanistic. Sometimes our mindreading explanations have the whiff of psychological essentialism, especially when we are efficiently and reflexively applying stereotypes to out-group members. Mindreading explanations involving generics introduce further variety into the taxonomy here. This diversity of explanatory practices in mindreading calls for a unified account of mindreading explanations. Such an account ought to be able to detail the kinds of explanations we employ in different contexts, explain whether this variety in explanatory practices is distinctive of the social realm, or whether similar patterns exist in other folk explanations, like perhaps folk biological explanations.

This is just a sampling of a few interesting, promising research projects based on the comprehensive view of mindreading developed in this book. I hope theorists in this field will adopt this broader view of mindreading that I argue for throughout this book. One need not accept my specific views, conclusions, or predictions to embrace the idea that mindreading is much more complex, diverse, messy, and interesting than the mindreading literature shows. After all, many mindreading theorists already acknowledge that the conception of mindreading used in the field is artificially narrow and restricted. As Ian Apperly remarks, "although it has long been recognized in principle that there should be important links between [Theory of Mind] and research on social psychology, reasoning, and experimental pragmatics, these literatures have seldom meshed well in practice" (2012, 837).[2] I hope that this book goes some way toward bridging these gaps.

Notes

1 Existing empirical work suggests that preschool-aged children conceptualize gender as a natural kind, but they do not typically do the same for race and other social categories. Later in childhood, children may come to essentialize more social groups or even view previously essentialized social groups as merely conventional. See Liberman, Woodward, and Kinzler (2017) for a review of these findings.

2 Hannes Rakoczy (2014) has made similar remarks.

References

Aboud, F. E. 2003. "The formation of in-group favoritism and out-group prejudice in young children: Are they distinct attitudes?" *Developmental Psychology* 39 (1): 48–60. doi:10.1037/0012-1649.39.1.48.

Abrams, D., and A. Rutland. 2008. "The development of subjective group dynamics." In *Intergroup Attitudes and Relations in Childhood Through Adulthood*, 47–65. Oxford: Oxford University Press.

Abrams, D., A. Rutland, S. B. Palmer, J. Pelletier, J. Ferrell, and S. Lee. 2014. "The role of cognitive abilities in children's inferences about social atypicality and peer exclusion and inclusion in intergroup contexts." *British Journal of Developmental Psychology* 32 (3): 233–247. doi:10.1111/bjdp.12034.

Apperly, I. A. 2012. "What is 'theory of mind'? Concepts, cognitive processes and individual differences." *The Quarterly Journal of Experimental Psychology* 65 (5): 825–839.

Buttelmann, D., N. Zmyj, M. Daum, and M. Carpenter. 2013. "Selective imitation of in-group over out-group members in 14-month-old infants." *Child Development* 84 (2): 422–428. doi:10.1111/j.1467-8624.2012.01860.x.

Dunham, Y., A. S. Baron, and S. Carey. 2011. "Consequences of 'minimal' group affiliations in children." *Child Development* 82 (3): 793–811. doi:10.1111/j.1467-8624.2011.01577.x.

Liberman, Z., A. L. Woodward, and K. D. Kinzler. 2017. "The origins of social categorization." *Trends in Cognitive Sciences* 21 (7): 556–568. doi:10.1016/j.tics.2017.04.004.

McLoughlin, N., S. P. Tipper, and H. Over. 2017. "Young children perceive less humanness in outgroup faces." *Developmental Science*, e12539–n/a. doi:10.1111/desc.12539.

Rakoczy, H. 2014. "What are the relations of thinking about groups and theory of mind?" *British Journal of Developmental Psychology* 32 (3): 255–256.

Westra, E. 2017. "Pragmatic development and the false belief task." *Review of Philosophy and Psychology* 8 (2): 235–257. doi:10.1007/s13164-016-0320-5.

Xiao, N. G., R. Wu, P. C. Quinn, S. Liu, K. S. Tummeltshammer, N. Z. Kirkham, L. Ge, O. Pascalis, and K. Lee. 2017. "Infants rely more on gaze cues from own-race than other-race adults for learning under uncertainty." *Child Development*.

Index

Note: Page numbers in bold indicate tables on the corresponding pages.

accuracy in mindreading 43–44
Actor-Observer Effect 57n11
Andrews, Kristin 16, 17
Apperly, Ian 101
approaches to mindreading 45–52
Arvan, Marcus 75n4
Autism Spectrum Disorder 5, 16, 20n2, 24, 67, 100

Ballantyne, Nathan 95–96n6
Bayesian models of learning 2
behavioral schemata 3, 16
behavioral scripts 34
belief attribution 23–25
bias: Actor-Observer Effect 57n11; centrally encoded 30; confirmation 50–51, 90; fighting 96n11; Group-Serving Attributional Bias 49–50; implicit 28–29, 35, 37n11; in-group/out-group 31–34, 36n1, 38n13, 49–50, 83–84, 94, 95n5, 96n9, 100; Naïve Realism 37n3, 50, 57n3, 82–83, 90; self-enhancing 84, 86–87; Self-Serving Attributional 49–50; social 27–34, **30–31**, 36n1
broad scope of mindreading claim 9

categorization, social 17–18, 25–26, 37n5
Clinton, Hillary 33, 38n13
cognition, embodied and enactive 9–15, 19, 20n8
cognitivism 9, 20n8
conceptually central associations 30–31, **31**
confirmation bias 50–51, 90
constructivist theories 2
context, situational 34–35
curse of knowledge 57n7

Deductive-Nomological model 15
Del Pinal, G. 29, 36n1
Dennett, Daniel 23
dependency networks 30
disagreement, peer 85–88
dual-process version of ST 67
Dunning-Kruger effect 82

efficiency, mindreading 45–46
embodied and enactive cognition (EC) 9–15, 19, 20n8
epistemic and ethical applications of mindreading: application to epistemology of peer disagreement 85–88, 95n4, 95n5; epistemic justice 81, 88–93, 96n7; isolation of mindreading literature and 80–81; judgments about others' knowledge and competence in 81–85
epistemic justice, interventions 81, 88–93
explanandum 24, 42, 55–56, 66
Explanation for Export hypothesis 54, 58n14
explanations 14–15; formal 52–56, 57–58n12, 58n14, 70, 100; mechanistic 52–56, 58n16, 70, 100; nature of 52, 54, 100–101; teleological 20n3, 52–56, 70, 100; types of 52–56, 58n16
explicit bias 29

false beliefs 23–24
false-belief task 23–24, 42, 71, 73
false consensus effect 57n7
folk psychology 1, 3, 69–70, 76n14; defined **8**; embodied and enactive cognition (EC) in 9–15; pluralistic 15–19; primary intersubjectivity in 10; regulative 16
formal explanations 52–55, 58n16, 100–101
Fricker, M. 89, 92
functional explanations 53–54
function of mindreading 56–57n2

Gallagher, Shaun 9–15
gaze cueing 99–100
generics 47–78, 100
goals of mindreading 43–44, 56, 57n4
Godfrey-Smith, Peter 68
Goldman, A. I. 67
Group-Serving Attributional Bias 49–50

Hillbilly Elegy 32
hiring processes 91–92
human beings as social creature 1

Implicit Association Test (IAT) 28–30, 38n14
implicit bias 28–29, 35, 37n11
implicit mindreading 14–15
in-groups 31–34, 36n1, 38n13, 83–84, 94, 95n5, 96n9, 100; Group-Serving Attributional Bias and 49–50
injustice, epistemic 81–82
intersubjectivity 9–10
isolation of mindreading literature 80–81

judgments about others' knowledge and competence 81–85
justice, epistemic 81, 88–93, 96n7

King, Nathan 86
Kuhn, T. S. 74–75n2

Lai, Calvin 37n11
Lombrozo, T. 58n13

Maibom, Heidi 68, 76n14
Malle, Bertram 58n16
McGeer, Victoria 16, 17
mechanistic causal explanations 58n16
Medina, J. 89, 92
mindmisreading 95n1
mindreading 1; approaches to 45–52; belief attribution and 23–25; bias in (*see* bias); broad scope of mindreading claim 9; conclusions on 98–101; defined 7, **8**; embodied and enactive cognition challenge in 9–15; errors in 48, 98; function of 56–57n2; general theories of 64–68, 74–75n2, 74n1; goals of 43–44, 56, 57n4; implicit 14–15; introduction to 7–9, **8**; isolation of literature on 80–81; narrow focus of standard theories of 23–25, 35–36; orthodox view of 2–3, 5n1, 20n4; products of 52–56; reasons for 42–43; situational context in 34–35; social categorization in 17–18, 25–26; standard terminology in 7, **8**; tacit 11–12, 14–15; understanding another person's behavior as 1, 5n2
mindshaping 16–17, 44
Model Theory 68–72, 75–76n12, 98–99; assessment of 72–74; epistemic justice and 88–89; function of theories and 62–64; general theories of mindreading and 64–68, 74–75n2, 74n1; as version of Theory Theory 71, 72

Naïve Realism 37n3, 50, 57n3, 82–83, 90
nativist theories 2
normative theories 62
norms, social 57–58n12

Obama, Barack 32–33, 38n13
out-groups 31–34, 36n1, 38n13, 83–84, 94, 95n5, 96n9, 100; Group-Serving Attributional Bias and 49–50

peer disagreement 85–88, 95n4
phenomenology 12–15, 19, 20n6, 20n7
philosophical theories vs. scientific theories 62–64
plausibility points 63
pluralistic folk psychology 15–19
prediction 14–15, 20n3
prejudice 33–34
Premack, David and Guy Woodruff 23
primary intersubjectivity 9–10

priming tasks 28–30
products, mindreading 52–56

reasons for mindreading 42–43
regulative folk psychology 16

salient-statistical associations 30, **30**
Schwitzgebel, E. 13
secondary intersubjectivity 9–10
self-enhancing bias 84, 86–87
Self-Serving Attributional Bias 49–50
Simulation Theory (ST) 2, 64–68, 71, 73–74, 75–76n12; belief attribution in 24; dual-process version of 67
situational context 34–35
social biases 27–34, **30–31**, 36n1
social categorization 17–18, 25–26, 37n5
social cognition **8**
social manipulation 56
social role concepts 57–58n12
Spaulding, S. 10, 13, 20n6, 29, 36n1, 75–76n12
stereotypes 17–18, 26–27, 35, 36n2, 37n4–6, 46–49; formal explanations and 53, 57n8, 57n10; warmth judgments in 95n2
stereotypical explanations 53, 55

tacit mindreading 11–12, 14–15
teleological explanations 53–55
theories: function of 62–64; general mindreading 64–68
theory of mind 1, **8**; Autism Spectrum Disorder and 5, 16, 20n2; false beliefs and 24
Theory Theory (TT) 1, 64–68, 71–72, 73–74; belief attribution in 24; *see also* Model Theory
Trump, Donald 33

values, group 32–33
Vance, J. D. 32

Westra, Evan 72
Wimmer, Hans and Josef Perner 23–24
world views, confirmation of 50–51

Zawidzki, Tad 16–17